Automated/ Mechanized Drilling and Countersinking of Airframes

Other SAE books of interest:

Care and Repair of Advanced Composites, Second Edition
By Keith B. Armstrong
(Product Code: R-336)

Aircraft Maintenance: The Art and Science of Keeping Aircraft Safe
By Bruce R. Aubin
(Product Code: T-115)

Mechanics Modeling of Sheet Metal Forming
By Jwo Pan and Sing C. Tang
(Product Code: R-321)

Automated/ Mechanized Drilling and Countersinking of Airframes

By George N. Bullen

SAE International

Warrendale, Pennsylvania
USA

SAE International
400 Commonwealth Drive
Warrendale, PA 15096-0001 USA

E-mail: CustomerService@sae.org
Phone: 877-606-7323 (inside USA and Canada)
724-776-4970 (outside USA)
Fax: 724-776-0790

ISBN 978-0-7680-7646-2
SAE Order Number R-416
DOI 10.4271/R-416

Library of Congress Cataloging-in-Publication Data

Bullen, George N.
Automated/mechanized drilling and countersinking of airframes / by George N. Bullen.
pages cm
Includes bibliographical references.
ISBN 978-0-7680-7646-2
1. Rivets and riveting, Aircraft--Automation. 2. Airframes--Design and construction--Automation. 3. Drilling and boring machinery--Automation. 4. Manufacturing processes--Automation. 5. Fasteners. I. Title.
TL671.5.B85 2013
629.134'2--dc23
2013002026

To purchase bulk quantities, please contact:
SAE Customer Service
Email: CustomerService@sae.org
Phone: 877-606-7323 *(inside USA and Canada)*
724-776-4970 *(outside USA)*
Fax: 724-776-0790

Visit the SAE International Bookstore at
books.sae.org

Dedication

To my wife, Natalie, for her support and patience,
and Joe Wardell, who was my right arm through
the long days and nights of design, development,
and transition to production.

Table of Contents

Preface xiii

Acknowledgments xv

Introduction 1

End Notes 16
Reference 16

Chapter 1 Historical Perspective: Why Airframes Have Holes . . 17

1.1 Introduction 17
1.2 The Airframe 18
1.3 Aluminum 19
1.4 Rivets 20
1.5 Assembly Tooling 23
1.6 Titanium 24
1.7 Hand Tools and Drills 25
1.8 Composites 28
1.9 Tooling and Composites 30
1.10 On-Assembly Drilling 32
References 36

Chapter 2 Airframe Manufacturing Cost Drivers 37

2.1 Introduction 37
2.2 Why Reduce Cost? 38
    2.2.1 The Value Price Profitability Index Commercial 40
2.3 Defining Cost 41
2.4 Identifying Manufacturing-Related Costs 43
    2.4.1 Efficiency 44
    2.4.2 Effectivity 46
    2.4.3 Impact 47
    2.4.4 Cost Contributors Assessment 51
2.5 Summary 52
References 54

Chapter 3 Incentives and Disincentives to Automate 55

3.1 Introduction 55
3.2 Military Automation Disincentive 57
3.2.1 The Means of Production 57
3.2.2 Parceled Application 59
3.2.3 Specific Use 60
3.2.4 Time Horizon 60
3.2.5 Politics 61
3.2.6 Controlled Profit 62
3.2.7 Restricted Sales 62
3.2.8 No Competition 63
3.3 Military Producers' Incentives to Automate 64
3.3.1 Affordability 64
3.3.2 Enlightenment 64
3.3.3 Complexity 65
3.3.4 Multiyear Buys 65
3.3.5 Quality 65
3.3.6 Marketing 66
3.3.7 Access 66
3.4 Commercial Disincentives 67
3.4.1 Geopolitical Local 67
3.4.2 Geopolitical Global 68
3.4.3 Labor 68
3.5 Commercial Incentives 69
3.6 Summary 70
References 71

Chapter 4 Types of Drill/Countersink Automation 73

4.1 Introduction 73
4.2 Motion 74
4.2.1 Size 75
4.2.2 Machine Types 76
4.2.3 Tools and Parts 80
4.2.4 Integration Level I 81
4.3 Action 81
4.3.1 End Effector 81
4.3.2 Dust Collection 84
4.3.3 Coolant 85
4.3.4 Tool Changer 85
4.3.5 Vision System 85
4.3.6 Inspection 85
4.3.7 Integration Level II 86

4.4 Controller 86
4.4.1 Human Machine Interface 86
4.4.2 Volumetric Compensation Algorithm 87
4.4.3 Integration Level III 87
4.5 Calibration 88
4.6 Sustainability 89
4.7 Summary 89
References 90

Chapter 5 Considerations Before Replacing Hand Drilling with Mechanization or Automation 91

5.1 Introduction 91
5.2 Before the Decision to Automate 93
5.2.1 Feasibility Analysis 93
5.2.2 Cost Benefits Analysis 100
5.3 "Go-Ahead" Decision Criteria 105
5.4 Summary 105

Chapter 6 How to Choose the Right Automation 107

6.1 Introduction 107
6.2 Three Don'ts 108
6.2.1 Needs 109
6.2.2 Affordability 110
6.2.3 Complexity 112
6.3 References 112
6.4 Requirements Document 113
6.4.1 General Information 114
6.4.2 System Summary 114
6.4.3 Functional Requirements 114
6.4.4 User Impacts 114
6.4.5 Performance Requirements 114
6.4.6 Additional System Requirements 115
6.4.7 Equipment and Software 115
6.4.8 Security 115
6.4.9 Location 115
6.5 Summary 116
References 117

Chapter 7 Steps to Installation 119

7.1 Introduction 119
7.2 Mapping the Existing Process 120
7.2.1 Part Placement 122
7.2.2 Trim 124
7.2.3 Finished Hole Size 124

7.3 Mapping the New Process 125
7.4 Requirements Document 126
7.4.1 Identify Stakeholders and Define Stakeholder Needs 126
7.4.2 Categorize System Requirements 127
7.4.3 Interpret and Record Requirements 128
7.4.4 Sign Off 128
7.5 Platform Specification 128
7.6 Request for Proposal 129
7.6.1 Introduction 130
7.7 Manufacturing Readiness Level Consideration 132
7.8 Performance Measurement Tools 133
7.9 Factory Acceptance 133
7.9.1 Basic Process Control System 134
7.9.2 Hazard 135
7.9.3 Instruments 135
7.9.4 Process Risk 135
7.9.5 Safety Instrument Function 136
7.10 Installation and Test 136
7.11 Summary 136
References 137

Chapter 8 Test Procedures 139

8.1 Introduction 139
8.2 Unique Tests 140
8.2.1 Impact Avoidance Test 141
8.2.2 Aluminum Tape Ink Dot Test 143
8.2.3 Compression Deflection Test 146
8.2.4 Drill Routine Conformance Test 147
8.2.5 Countersink Depth Test 147
8.2.6 Machine Subroutines 148
8.2.7 Final Full Function 149
8.2.8 Demonstration 149
8.3 Volatile Organic Compounds 150
8.4 Summary 151
References 152

Chapter 9 Transition to Production 153

9.1 Introduction 153
9.2 TTP Document and Transfer of RAA 154
9.3 TTP Technology Readiness Levels 155
9.3.1 TRL 4 155
9.3.2 TRL 5 156
9.3.3 Combined TRL/MRL 156

9.4 TTP Manufacturing Readiness Level . . . 156
9.4.1 MRL 4 Material Solutions Analysis, Milestone A . . . 157
9.4.2 MRL 5 Technology Development, Milestone B . . . 157
9.5 TTP and Product Delivery Team Composition and Selection . . . 158
9.6 TTP Document . . . 159
9.6.1 TTP Final Hole Count and Cost Benefits Analysis . . . 161
9.7 TTP Risk Identification and Assessment . . . 162
9.7.1 Technology and Industrial Base . . . 163
9.8 Validation of System to Production Requirements: Exit Criteria . . . 166
9.9 Summary . . . 166
References . . . 167

Chapter 10 Training . . . 169

10.1 Introduction . . . 169
10.2 Who to Train . . . 171
10.2.1 Operator . . . 171
10.2.2 Design Engineer . . . 177
10.2.3 Manufacturing Engineer . . . 180
10.2.4 Quality Assurance . . . 182
10.2.5 NC Programmers . . . 183
10.2.6 Cost Center Managers . . . 183
10.2.7 Tooling . . . 184
10.3 Summary . . . 184
References . . . 185

Chapter 11 Maintenance . . . 187

11.1 Introduction . . . 187
11.2 Complexity—New Normal . . . 188
11.3 Impact Prevention . . . 191
11.4 Predictive Maintenance . . . 192
11.4.1 Maintenance Integration . . . 192
11.5 Prognostic Health Monitoring . . . 196
11.6 Maintenance . . . 196
11.6.1 Autonomous . . . 197
11.6.2 Breakdown . . . 198
11.6.3 Improvement . . . 198
11.6.4 Periodic Maintenance . . . 198
11.6.5 Certification . . . 198
11.7 Buying Maintenance . . . 198
11.8 Troubleshooting . . . 199
11.9 Summary . . . 200
References . . . 201

Chapter 12 Social Impact of Automation 203

12.1 Introduction 203
12.2 Pacifying the Human Capability-Machine Experience 204
12.2.1 The Social Concept of Operations 206
12.2.2 Data Compilation 212
12.3 Perception of Ubiquitous Automation 212
12.4 Safe, Secure, and Ethical Automation 213
12.4.1 Safety 213
12.4.2 Security 214
12.4.3 Ethical Application of Automation 214
12.5 Management of Automated Systems 215
12.6 Organizations 215
12.7 Customers 216
12.8 Summary 216
12.8.1 Communication 217
12.8.2 Participation 217
12.8.3 Consultation 217
12.8.4 Training 217
12.8.5 Support 217
References 218

Chapter 13 Future State of Airframes and Automation Summary 219

13.1 Introduction 219
13.2 Economics 220
13.3 Manufacturing Exodus 221
13.4 Exit the Autoclave 225
13.5 Automation or Labor 227
13.6 Fastener Alternatives 229
13.7 Summary 230
References 233

Appendix A Automated Machine Suppliers for D/C of Airframes 235

Index 237

About the Author 245

Preface

This book has been written to guide airframe producers through the challenging process of applying automation to perform drilling and countersinking (D/C). Automated D/C is the application of specialized Numeric Control (NC) machinery to D/C a finished hole "one shot"[1] for insertion of the fasteners that hold the airframe together. Applying automation replaces the traditional hand methods for D/C. A chapter-by-chapter description will be provided to define the steps necessary to successfully apply automation for D/C holes.

There can be well over 100,000 holes that need to be drilled and countersunk in a modern military fighter/attack airframe and more than 2,000,000 in a large commercial airframe. Large commercial and military airframes can exceed three million holes. The large number of holes and the hand method for D/C drives cost, quality, schedule, complexity, and safety. Sixty-five percent of the cost of airframe assembly is derived from D/C of holes by hand. Eighty-five percent of the quality issues and eighty percent of the lost time injuries are the result of the hand process for D/C. Therefore, the focus of this book will be the area of greatest return on investment (ROI) for the airframe producer—automation of D/C.

Numerous attempts have been made to automate the fastener installation and sealing process, with limited success. The investment, complexity, processes, and operation of fasteners have not made their application economically viable as an option to replace the hand operation, in most cases. Examples exist, but they are limited to off-line assembly operations or are rare in on-assembly applications.

There are some signs that both commercial and especially military aircraft have reached a price point that is no longer affordable. Two examples are the F-22 that was canceled in 2011 as too costly and the F-35 that has grown in cost to the extent that the U.S. Government has told Lockheed Martin Corporation to absorb any additional cost overruns. Delays and cost overruns have postponed the release of the Airbus A380 and Boeing 787 Dreamliner.

Automated D/C is a proven cost reducing technology that improves quality where it has been applied successfully.

1. "One shot" means drilling a finished hole in operation without pilot holes or tool changes. The hand method of drilling and countersinking requires multiple steps and tool changes.

Acknowledgments

The author would like to acknowledge the contribution of those in leadership, teammates, colleagues, and friends at:

Northrop Grumman Corporation
El Segundo, California

Northrop Grumman Corporation
Palmdale, California

Northrop Grumman Corporation
Huntsville, Alabama

Lockheed Martin
Fort Worth, Texas

M Torres
Pamplona, Spain

AIM
San Dimas, California

Purdue University
West Lafayette, Indiana

California State University
Long Beach, California

Oxford University
Oxford, UK

University of California at Los Angeles
Los Angeles, California

Loyola Marymount University
Los Angeles, California

Pepperdine University
Malibu, California

Advanced Manufacturing Research Center
Sheffield, UK

Introduction

So what has changed?

—Richard C. Bullen

My father retired from the U.S. Air Force in 1962, packed up the family, and moved to California where he went to work for Northrop Aircraft as an engineer. In 1968 he moved to Bremen, Germany where he finished a long career in the U.S. Air Force and aerospace at VFW Fokker. In 1996, just before his 77th birthday he asked if he could come to California and see, "That new F/A-18 E/F production line." I was working for Northrop Grumman Corporation in El Segundo, California where the new F/A-18 E/F production line was in construction across the aisle from the old F/A-18 C/D assembly line. So I acquired the necessary signatures and proudly walked the assembly line with him. At the end of the assembly line, he turned and looked first to one side and then the other. First at the C/D line completed in the seventies and then at the E/F line just being finished.[1] After a few moments of reflection, he looked at me and asked, "So what has changed?"

In fact, he must have thought we were going backward. The F-5 Northrop Overhead Rail Assembly and Installation Line (NORAIL) system was producing aircraft when he was at Northrop's Hawthorne facility.[i] Built in the 1950s, it was a revolutionary concept based on a rolling line without the traditional boxed assembly jigs or numerous crane operations of today's assembly lines.

1. It is not unusual in aerospace to have a new assembly line installed next to a mature or dying assembly line. The life span of an aerospace assembly line can exceed 25 or 30 years. Building a new line next to an older line facilitates the transfer of skilled mechanics and maximizes the use of shared resources.

Since that day much has changed. If my father were alive today, the changes are overt enough that he would notice them. However, he would have to go to the right places.

Changing processes and technology has been painful and isolated to a handful of companies incentivized for a variety of reasons.[2] Access to structure is one incentive, as illustrated in Fig. I.1.

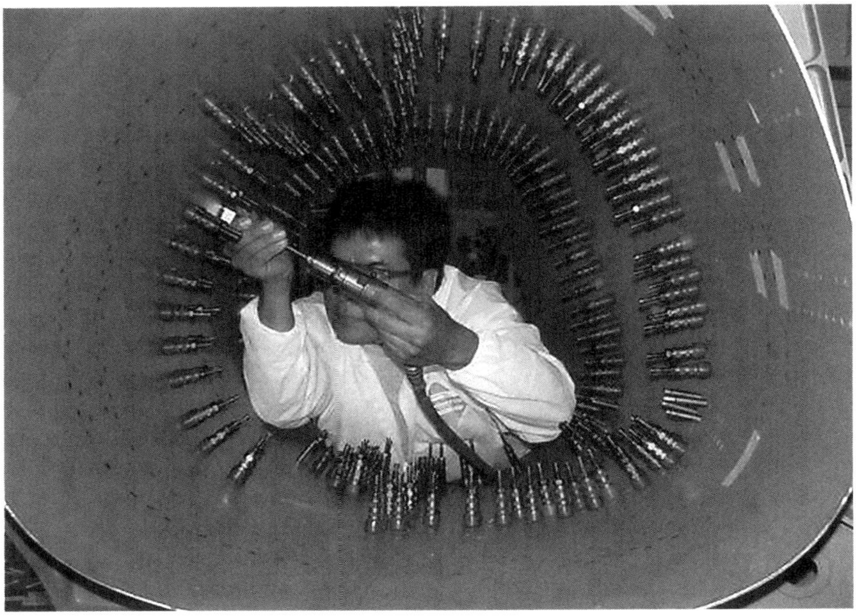

Fig. I.1 Difficulty accessing structure.

The most obvious and visible change has been the incorporation of on-assembly automated/mechanized D/C technology. The change from hand to automated/mechanized D/C has also had the largest affect on decreased cost, increased quality, and decreased lost time injuries. The U.S. Government refers to on-assembly automated/mechanized D/C technology as a *Disruptive Technology*.[ii] It is disruptive because of its positive effect on the cost and quality of the products where it is applied. It is also disruptive because of the trauma it causes to the companies, people, processes, and factories where it is applied. Airframes are complex products produced in specialized factories by highly skilled people. The processes and technologies that have led to production of airframes have evolved, are understood, and robust to the extent that the hand manufacture of a product can be. Change to a successful decades old, albeit costly manufacturing method is difficult. The content of this book is the result of lessons learned from the painful experience of

2. Tighter tolerances that have challenged the traditional hand method of drilling have also incentivized companies to look to automated D/C.

a thousand mistakes that led to successful installation of on-assembly automated/mechanized D/C technology around the world.

Throughout this book the terms *automated* and *mechanized* will be referred to as *automated*. There are large and small differences between on-assembly automated systems and on-assembly mechanized systems. The primary difference is that automated systems are large and derive their motion from controllers such as the Siemens 840D[3] that are made for complex, multi-axis systems. Smaller systems that have three axes of motion and sometimes two axes of motion and are driven by a PLC (Programmable Logic Controller) are mechanized systems.[4] The abbreviated terms are monikers in the aerospace and aerospace machine manufacturing industry and are used interchangeably.[5] Where it is notable or relevant, the system type will be designated as peculiar to automated or mechanized.

Chapter 1 provides a historic perspective of how aerospace evolved into its current state. Drilling and filling are the primary activities of airframe assembly. Because the fastener follows the hole, D/C is critical to the performance and safety of an aircraft. Large infrastructures and complex processes to support the tools and machines that D/C have evolved and are in place today. It is important to know how aerospace arrived at its present state to understand the variety of activities that support the very complex process of drilling and filling. By understanding how aerospace got to where it is today, it becomes easier to understand and project where aerospace will be tomorrow and what industry, governments, academia, and professional societies need to do to help them get there.

Chapter 2 answers the question, "Why choose D/C?" If change in any manufacturing process is made based on its positive effect on the product, the greatest ROI in change should be realized. D/C is directly and indirectly the largest cost driver for assembling an airframe. The costs to D/C are a big target for maximizing the effect to incentivize change from hand to automated processes. The engine that drives the manufacturing process is composed of the elements cost, quality, and schedule. On-assembly automated D/C offers the largest and most visible effect for positive performance improvement to the elements of the manufacturing process engine. Sixty-five percent of the cost to assemble the airframe is directly related to the labor effort to D/C holes. Sixty-five percent of the entire aircraft cost is attributable to the airframe. If a disruptive technology is to be introduced into the manufacturing stream, the largest target is the easiest to hit. Direct ROI business models are discussed in this chapter to provide the necessary tools for the reader to make reasonable assessments for automating

3. I have used the Siemens controller as an example and not as a recommendation. They are a fine controller but there are other controllers available that may meet the specific application needs of the automation applications engineer.

4. Embedded, highly specific PLCs are now emerging that offer lower cost and simpler programming.

5. The aerospace industry identifies airframe components, parts, and tooling by acronyms and monikers. For example, a lifting and carrying fixture for the F-18 is called "Elephant Ears" due to its appearance.

the D/C process. The ancillary benefits to change from hand to automated D/C are also critical and discussed. Although it is important to remember that an ROI should be based on direct cost benefits analysis, other contributors to produce the product such as health and safety, quality, and assembly line continuity are also affected. It is also important to remember the cause and effect rule. The positive aspects of automated D/C can be offset by increased cost in other areas that must be considered in the decision model to change from hand D/C to automated D/C. This is not a book about the positive aspects of automated D/C, but a book that provides at its core both positive and offsetting considerations that can lead to either a go or no-go decision for change. For example, the automated drilling of holes results in a consistent/repeatable hole very close to the drill bit size. The hand drill method makes "looser" holes[6] so that acceptable fasteners throughout the tolerance band will fit. As a result of automating D/C, some of the fasteners may not fit into the hole. Therefore, an added cost to procure tighter-tolerance fasteners or select specific fasteners to fit the hole must be amortized into the cost of automation. This is just one example that must be factored into any decision to automate. It is good to improve quality and sing the mantra, pursuit of perfection. However, consideration must also be given to business operating margins and the importance of offsetting factors that can drive the cost of automating D/C outside reasonable and acceptable boundaries.

Chapter 3, Incentives and Disincentives to Automate, addresses many of the reasons that drive companies to consider automation and either accept or reject it. There are many considerations that can determine the final decision to begin the long and arduous road to automate D/C. In one case, I was told, "I have a mature production line. The customer is happy with cost, quality, and performance. Why would I want to change anything?" Case closed.

In another case, the product had a rate of five airframes a year and a very low labor rate. The ROI would be impossible to meet within the product lifecycle. This particular application case was especially disappointing from an automation D/C perspective. An airframe that has long rows of holes of the same size through very thin sheet metal with unobstructed access to the airframe is an automated D/C application engineer's dream. But the ROI just was not there.

However, both the described application analysis cases were for mature production lines past the peak of their lifecycles. Of course the best application is at the beginning of the life cycle of a product when the production line is being planned. However, there are also opportunities available on existing production lines. Both have opportunities to invest and realize the returns that can incentivize a decision to automate D/C. Consistent quality, highly predictable process times, and reduced cost are just a few.

There are a variety of drill/countersinking automated systems available and described in Chapter 4. The types of systems for on-assembly automated D/C

6. Meets acceptable specified limits.

are unlimited because on-assembly automated systems are custom designed and built for specific applications. On-assembly automated D/C machines are not Commercial Off-The-Shelf (COTS) systems.[7] They are not "catalog" machines such as Haas milling machines[8] where a buyer goes with a specification in hand to a showroom or machine show. Very few people design and make on-assembly automated D/C machines. I have included a list of the more prominent manufacturers of on-assembly D/C machines at the end of the book in Appendix A. Properly selected and judicially applied, on-assembly automated D/C is essential to a good product. In the airframe assembly world, structure is located in a large rigid assembly fixture. Inside the assembly fixture the structure is held in location with a limited combination of temporary fiduciary fastening devices such as clamps, clecos, and temporary undersize fasteners. The structure is referred to as green until most or all of the permanent fasteners are installed, and the subassembly can stand alone outside the assembly fixture. Drilling the green structure in place with aids contributing and supporting all the pieces of the subassembly facilitates conformance to specified tolerances and dimensional stability of the detail components. Conversely, if a subassembly is moved after temporarily fastening the pieces together, an unnecessarily large risk of the components shifting and violating the dimensional integrity of the product is introduced. Drilling and filling in a green condition can cause assembly variance and affect downstream joining and fit up.

Putting a machine into the assembly cell is the optimal place to perform the drilling task. On-assembly automated D/C replicates the best aspects of the hand drill process where the structure is most stable.

Chapter 5, Considerations Before Replacing Hand Drilling with Automation or Mechanization, presents the variety of decision models available to identify opportunities for on-assembly automated D/C. There are obvious benefits to the on-assembly automated D/C opportunities that exist when designing an assembly line from scratch. A different decision model must be used for an existing assembly line, including the considerations for changeover from hand to machine operation. Not all opportunities in either application are necessarily suited for both drilling and countersinking, or drilling full size in one shot.[9] Sometimes a process such as

---

7. This is not to say that many of the components of on-assembly automated D/C machines are not COTS.

8. Haas automation produces a very good set of NC machines that are the workhorse of small, medium, and large machine shops. They are mass produced at the Haas Automation factory in Oxnard, California.

9. The ideal D/C operation is to drill "one shot" through both skin and substructure. One shot is defined as one pass without return to the hole for secondary enlargement of the hole or countersink. The hole is finished in "one shot."

cold working[10] prohibits the full-size drilling of holes. Some holes do not require countersinking or are in areas that prohibit the access of a D/C end effector. One tool provided in Chapter 5 is a Microsoft Excel decision matrix that will help calculate the intersection of the marginal return sloped line with the marginal cost sloped line. Above the intersection of the two lines (equilibrium), the ROI begins to increase. Below the equilibrium point, the ROI will be negative. This matrix and decision model is based on a number of factors including access, hole quantity, and size. It should be noted that ROI is not and should not be the only criteria for deciding to install any form of automation, including on-assembly automated D/C. It is, however, a business baseline from which all the other considerations flow.

At an automated fastening conference I listened to a speaker say he would not be satisfied until every hole on every airplane was automated. I am not of that opinion, and this book approaches the automation of D/C from a practical, low-hanging fruit perspective. There are holes in the airframe that would be prohibitively difficult and expensive to apply automation for D/C. Some things humans do better than machines. Some things humans do not do as well as machines, but they do them more efficiently at an acceptable quality level. Airframes are fastened together by tens of thousands of fasteners, and there are plenty of reasonable opportunities for application of on-assembly automated D/C without taking on extreme challenges or trying to compete with low labor rates.

How to Choose the Right Automation (Chapter 6) provides a common sense approach to automation. Monuments abound providing testimony to requirements creep and elevated expectations. I assure you that it need not be that way. *Keep it simple* and *take baby steps* are phrases that apply here and throughout this book. The pathway to automation failure is paved with overaggressive approaches to applying technology. If all you do is D/C holes, then the payback will be well worth the investment. In many cases, if all you do is drill holes, the payback in cost savings and quality will justify your investment. Choosing the right complexity of technology for the specific application of drilling a specific type of hole is the key to success.[11] Remember that the machine will do one thing over and over for its entire life cycle. I have had many an argument with people who try to put on-assembly automation D/C in the same category as machine tools. Perhaps they share some DNA, but they are different in that the task assigned to automated D/C makes repeatability more important than accuracy. If the machine will drill a thousand holes the same point every time, and the points can be mapped and brought into conformance, then the price differential does not warrant paying for

10. Cold working is a process where a hole is drilled undersized and subsequently a device inserted into the hole to compress the inside diameter (ID). This operation work hardens the hole and extends the durable life of the hole. A secondary operation is done using a reamer to remove a small amount of material to bring the hole up to the specified diameter but leaving a hardened amount of material on the inside. The cold working process has been automated using an expanding mandrel.

11. My father used to say, "You don't need an 18 wheeler to drive the kids to school."

accuracy. These and other considerations such as incremental indexing[12] can reduce the cost of a piece of on-assembly automation by half, with maintenance and spares costs reduced as well. Masses of steel and 20-foot-deep foundations with isolation insulation limit the reconfigurability of your factory for future optimization (if it is necessary). Monuments add nothing to the prestige of a factory.

The installation steps necessary to bring a piece of on-assembly automated D/C online are detailed in Chapter 7. Some are linear by necessity and others can safely proceed concurrently. A critical path analytical model is provided. Again, these are, to a large extent, custom pieces of machinery where there are some unknowns that have to be planned and backup plans (as in, "go to plan B") established. Risk reduction models such as the one in Fig. I.2 can help to identify possible activities and tasks to assuage the concerns of the project manager.

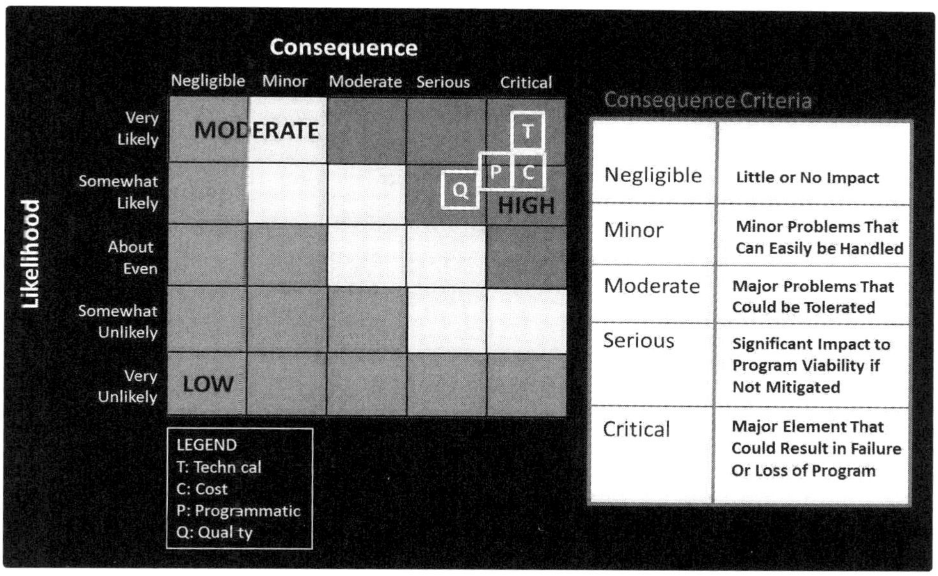

Fig. I.2 Risk assessment cube with consequence criteria.

While I believe it is necessary to actively engage in risk reduction activities, a measured approach to concern should also be exercised. Just as anxiety and concern can stifle a person to inactivity, so can unreasonable risk aversion relegate an on-assembly automated D/C project to a backroom science experiment or project. There will be many people who do not want change. From the top down, risk conjures fears driven by misconception that can result in negative consequence. A common social model is 5% for; 5% against; 90% on the fence. It is imperative that a precursor to the requirements document for the machine be the socialization and minimization of unwarranted fears with upper management. The leadership of the

12. Incremental indexing is the process of re-zeroing a machine to hard points along its axis of motion. Progressive degradation of the accuracy is mitigated by a circumvalence of hard index points.

company has to be fully aware of the risk and your ability to mitigate the risk. Every step of the installation process will be preceeded by a speed bump placed by the 5% against in the interest of their "concerns." The roadmaps provided in Chapter 7 will help identify and address these "concerns," no matter how trivial. The best defense against agents who resist change are constant vigilance, communication, movement forward, and the support of your leadership.

Test procedures described in Chapter 8 are an area where the artisans of speed bumps have honed their craft and elevated it to perfection. Let me be perfectly clear: do not lose sight of the goal. There is one thing and only one thing your fine piece of machinery must do: D/C holes where they are supposed to be within the diameter prescribed for the hole size. I have seen machinery shut down for weeks during the test and acceptance phase while gangs of experts armed with laser measurement devices combed every tenth of a millimeter of the machine envelop to laser map accuracy within ten millionths of an inch. Acceptance test procedures and the factory acceptance procedures are provided in this book. They will walk you through a detailed set of steps to ensure that before the drill tip hits the product, your machine will D/C good holes.[13] There are ways to ensure the attainment of well-drilled and -placed holes. Confidence and acceptance can be attained without the complex tools, technology, and procedures that exist and are necessary to calibrate, certify, and accept multipurpose machines.

Transition to Production (TTP) is where the testing ends and acceptance by the production team takes place. It is a very difficult part of the evolution of the machine application process when it is determined that it is ready to touch the product, especially if the production team either has never had experience with anything but the hand method of drilling holes. Inclusion and integration of the production team into the automation decisions and development from the start of the process helps alleviate their fears. The closer it comes to taking responsibility for running the automation, the more an eager production team will be to resist that final step. Figure I.3 graphically shows the relationships between the phases for bringing online an on-assembly automated D/C machine. The illustration also shows the relationship between the three phases and the Manufacturing Readiness Level (MRL).

Many large companies employ TTP teams that are skilled at bringing machinery online. They are usually drawn from the production floor and the development team. Figure I.4 shows the MRL relationship to the three phases of successful implementation.

TTP is also the linchpin that binds the cycle of successful development through production into a continuous circle of improvement. I have placed a number of tools and checklists in Chapter 9 for smoothing the transition to production effort.

---

13. There used to be, and possibly still is, a stone arch above the entry to the Ohio Trist Drill Factory. Inscribed in it are the words, *Our customers don't want drills! They want holes.* Much the same with on-assembly automated D/C. Good holes are the goal, not machines.

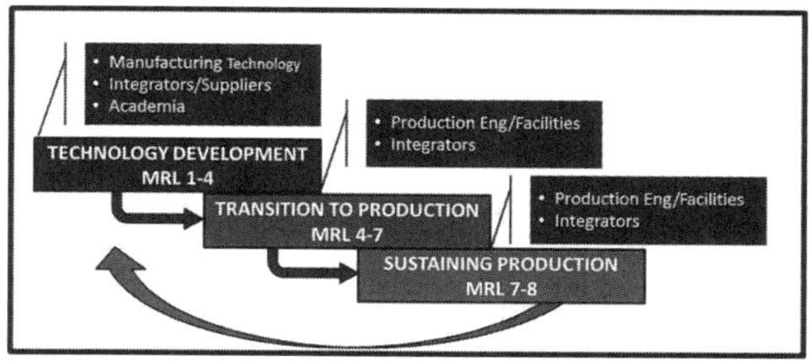

Fig. I.3 Three phases of development and MRL relationship.

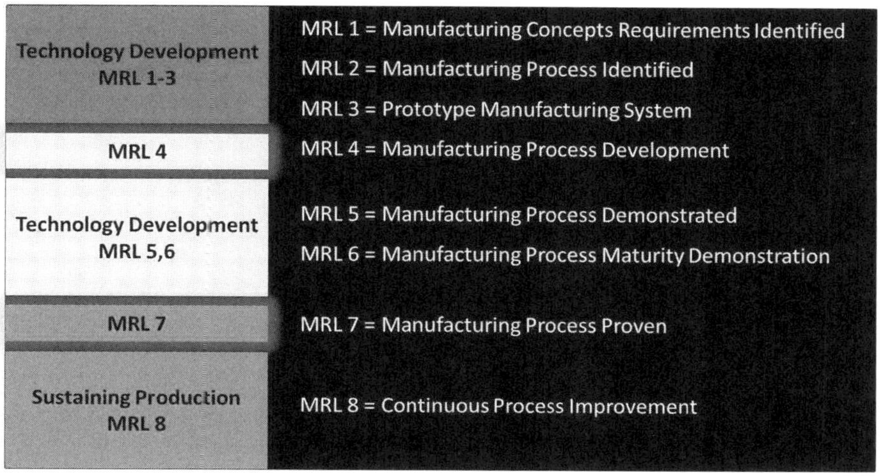

Fig. I.4 Three-phase implementation relationship to MRL.

Training is addressed in Chapter 10. It is important to remember that training is not limited to the operators of the machinery. All the people who make up the elements of the automation production team must be trained and in certain cases certified initially and annually. The people who need to be trained include leadership (including VP and director level), engineering (ME, IE, and design and stress engineers), facilities personnel, operators, and mechanics surrounding the work cell; and janitors. It is important to remember the change taking place and how it will affect every aspect of operations. Chapter 10 will provide a training plan with a matrix that identifies the individual job code, level, and type of training needed. With proper training, each participant in the automation chain will understand the differences between hand and machine D/C and manage the interaction between the two. In an aerospace environment, product will flow across bridges to islands of automation from islands of hand operations. How well the bridges function to coordinate the flow of data, information, resources, and product depends on the level of understanding that the entire production team has of the new integrated system. Training is a prime facilitator for the success of these dissimilar types of airframe assembly. Without proper training the bridges can become bottlenecks or even barriers.

Before any training takes place, a selection decision must be made about who will run the machine. While this chapter offers a good fundamental approach to who, what, and how to train and prepare the production team for on-assembly automated D/C, the hardest precursor to training the operator is selection. The success of the training and the operation of the machine depend on the right people running the machine. There are variables both controllable and uncontrollable that can affect selection of the automation operator. Uncontrollable variables include whether the factory is a union or nonunion shop[14] and increased wage compensation to incentivize applicants.

An important controllable variable is from what discipline the machine operator will be drawn. Two of the options are either an airframe assembly mechanic or a machine operator. This is a difficult decision that I have seen go either way.

The argument for a mechanic to become the operator stems from his or her intimate understanding of the airframe and the D/C process. The on-assembly automated system is a repetitive operation with a purpose of limited scope; the mechanic brings airframe experience to the automated operation. He or she is familiar with the specifications for good holes and their placement. This is a strong argument. When automation "goes bad,"[15] a mechanic can recognize any number of signs derived from years of familiarity with the process.

On the other hand, an NC machine operator brings familiarity with NC machinery that a mechanic probably does not have and can recognize a machine anomaly faster than a mechanic.

So the question is, does a better operator come from the mechanic in the assembly center, also known as the cost center, with years of airframe D/C experience, or does the better operator come from a machine operator trained to understand the airframe D/C process?

The edge goes to the airframe assembly mechanic with some caveats. My preference for the mechanic to be trained as the on-assembly automated D/C operator is due to their intimate knowledge of the airframe. The mechanic in many cases knows the airframe subassembly in their cost center better than anyone else and can determine if an error has or is about to occur.

The other reason is that the cost center teams are very close-knit groups of individuals who have spent long hours developing a rapport. By selecting a mechanic from the labor pool within the cost center, the team is more apt to accept the new technology and work together to ensure its success.

The successful insertion of new technology has as much to do with people as it does the technology capability itself. If I had to choose, the support of leadership and

14. A strong union can have an effect on the selection process. The development of a job description will have to be accomplished as part of the selection process that can be influenced by union input.

15. In the hand D/C operation, a mechanic drills a bad hole and stops after discovery of the error. A machine drills holes very quickly and requires quick response to limit the effect of errors.

the mechanics in the cost center (top floor and shop floor) are the critical human components that can make or break new technology transition to production. Training and inclusion are two key components to their support.

When new technology is introduced, it is essential that maintenance issues and their remedies as described in Chapter 11 are addressed. The realistic identification of spares and their acquisition, cataloging, and storage are areas often missed by the initial capital cost/benefits team in their calculations for ROI. It is an area that becomes a point of contention when the next machine goes online. It does not help a new technology migrate across the production line when capital budgets that are already established have to be reorganized and funding shifted from another acquisition to support the unforeseen need for a part. A recommended generic spares list is included in Chapter 11 as a baseline and reminder.

Planning for calibration is another area often forgotten in the initial frenzy to capitalize, acquire, and transition to production of a new automated D/C machine. Two areas, besides others described in this chapter, are timing to align calibration with the rhythm of the production line and specialized calibration equipment.

Airframe assembly lines normally have an annual shutdown to calibrate and perform maintenance on the assembly line.[iii] Airframe assembly lines are progressive,[16] and any disruption to the line causes costly schedule and line flow trauma. Therefore, the initial calibration schedules need to be harmonized with the rhythm of the rest of the assembly lines' annual shutdown.

Many of today's highly precise machinery use laser calibration systems such as those produced by Leica and Faro laser trackers. Figure I.5 shows the setup of one type of laser tracker system.

Initial test and certification as well as annual certification use these types of systems to calibrate the on-assembly automated D/C technology. If your company has a laser tracker system or systems, an evaluation should be done to see if the quantity available is enough to support the additional task. If another is needed, this is an additional cost to the system.

One alternative to buying a system is to either lease a laser tracker system when needed or lease a laser tracker team.

Chapter 12, Social Impact of Automation, was intentionally placed near the end and is meant to leave the reader pondering this important consideration before automating D/C. Installing automated D/C is a risk complicated by social issues. Automation of D/C reduces an existing workforce or reduces the projected numbers of workers to be hired. A positive ROI and the resultant decision to automate mean that a machine will replace people. Hours to produce holes have a direct

---

16. Progressive in the sense that each subassembly relies on all the other subassemblies to move in rhythm. If one subassembly is late to move, all the other subassemblies upstream are stalled. Subsequently, downstream subassemblies finish their work and become starved.

relationship to money. The only way to save enough to apply/install automation is to reduce hours consumed by labor to perform the same activity. A negative social perception by a workforce about automated D/C can defeat the successful application of automation before it begins or even worse, after the money has been spent to acquire and install the automated system. As technically oriented people we often minimize, ignore, or even mock the social considerations of highly technical decisions. The right socialization and approach, incorporating shaping of the perceptions of the workforce, are as integral to the speed of insertion and effective operation of the system as any other component.

Fig. I.5 Laser tracker system.

In addition to the mechanics, another social resistance area that has affected the transition to production of automated D/C is derived from the first- and second-level cost center managers. Two reasons that supervisors push back are budget reduction and reduced flexibility. The cost center's budget is reduced based on projected savings from automation while introducing a more complex operating system that the supervisor must manage.

Flexibility is reduced because automated D/C does one task and only one task very well. On the other hand, supervisors can move people from task to task, allowing them to shift labor where it is needed, as it is needed. Introducing a machine into the process reduces the mechanics in the cost center available to the cost center manager to adjust and focus labor where he or she may need extra effort.

These issues can be reduced or mitigated by a variety of methods. Chapter 12 describes many successful methods that have been used to help the workforce and their management adjust and accept automated D/C.

One additional item merits mentioning: if you or your company should decide to automate the D/C process, a strange transformation will occur. It is true whether you are automating the D/C process on an existing assembly line or a new assembly line. When the process is a hand D/C operation, it is assumed that if the hole has the proper edge distance (ED),[iv] then the hole is properly placed. Figure I.6 shows critical measurements for hole placement in the hand operation.

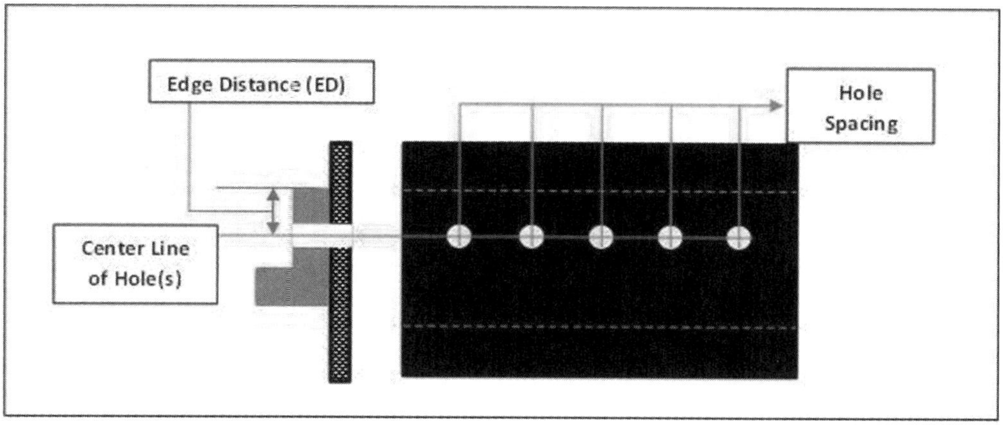

Fig. I.6 Critical measurements for hole placement in the hand drill operation.

The variables such as trim and part placement are critical to the hole's proper location and, therefore, ED.

The mechanic has at his or her disposal many tools to ensure ED. The tools and skills in the mechanic's possession to ensure proper edge distance may or may not derive from good trim or part placement. If the hole's ED is good, it is assumed the part was properly placed and the edge of the part was trimmed properly. In addition, if there is a violation to the tolerance control limits for ED and a defect occurs, the hole is tagged for cause and corrective action. The mechanic is coached as the process continues. More will be written about assumptions built into the hand D/C process in Chapters 7, 8, 9, and 10.

When automated D/C is installed, the machine goes through extensive calibration and acceptance tests. However, if the machine drills a hole that violates the control limits and results in a defect, all work stops. The machine is the assumed culprit. This is due to decades of acculturation convincing the manufacturing team that the upstream controls have worked. The confidence of a working system of controls is falsely validated to the manufacturing team by the mechanic consistently drilling holes with good ED.[17]

The assumption is that once the subassembly has reached the assembly floor, all its components are fabricated and placed in conformance to dimensional specification. Therefore, any defect has to be the result of what has changed. What has changed is the insertion of automated D/C. If the machine is found to be in conformance, which it almost always is, a dilemma confronts the manufacturing team. All the

17. Mechanics step drill holes beginning with a small pilot drill. After checking the ED of the pilot hole, the mechanic can "pull" the hole through progressively larger drills if they find the ED is out of conformance.

manufacturing processes performed up to the D/C operation are suspect and subject to review.

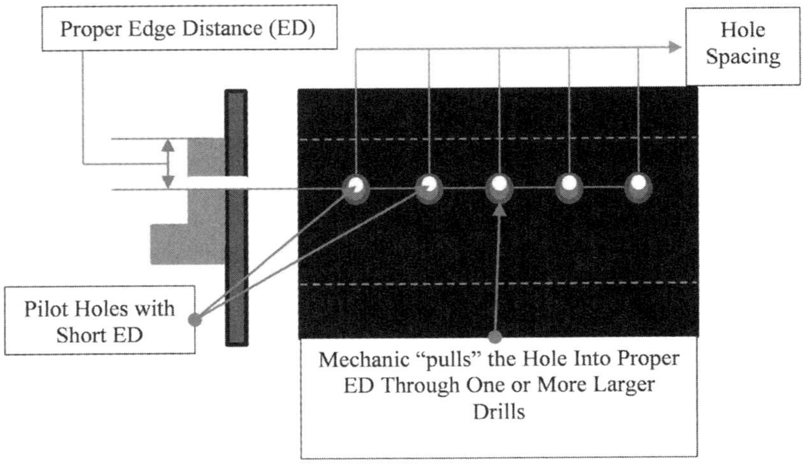

Fig. I.7 Mechanic "pulls" hole into conforming ED.

Discovery of an out-of-conformance process upstream of the automation reveals hidden costs and benefits of the automated D/C process. It is a cost because a more precise and expensive method of trimming and placing parts must be installed to run the automation. It is a benefit because a process that was assumed to be in control is unmasked to reflect reality and, therefore, is addressed and corrected. The result is a higher-quality product. Good automation will unmask the heretofore hidden areas of a process not in conformance.[18]

Chapter 13 concludes the book with a projection toward the inevitable, the future state of airframes and automation. Upon completion of this book, three things should be obvious. There are many fasteners in airframes. Fasteners and the holes needed for them are expensive, directly and indirectly. Using machines to D/C holes in certain applications is a better way to D/C holes than hand drilling, which is expensive and sensitive to human variability.

These three reasons are why research is being diverted to the eventual reality[19] that the automobile manufacturers arrived at decades ago. Reducing part count by unifying structure increases quality and decreases cost. Bonded structure is returning as a viable option to reduce fasteners and as an extension the associated reduction in D/C. Before anyone wants to stop and not read any further, let me say it is a long way off. Fasteners will be with us for a long time. As I stated earlier, aerospace is a conservative environment in which change comes gradually, and

18. In many cases after investigation, especially where parts are still hand trimmed, the cumulative tolerance for trim and part placement exceeded the allowable diametric tolerance for hole placement.

19. In Detroit it was more of an epiphany driven by panic derived from a loss of market share to Japanese auto manufacturers who were producing a superior product.

rightfully so. Airplanes are complex pieces of machinery that require considerable test and evaluation before change is introduced to ensure the performance and safety of the people who fly in them. However, there are changes coming that have been the result of new materials and the high cost of all those fasteners. In the beginning the Wright brothers built the first airframes from spruce and ash covered with muslin. Cloth-covered airframes continued to be the norm through World War I. Metal, specifically aluminum, migrated into the airframe as the principal material as it offered a stronger alternative to ever-increasing power and speed of aircraft that added stress to the airframe. Bonded structure returned in the form of the British de Havilland DH.98, the fast, furious, and stealthy Mosquito. Figure I.8 shows this agile aircraft of bonded structure.

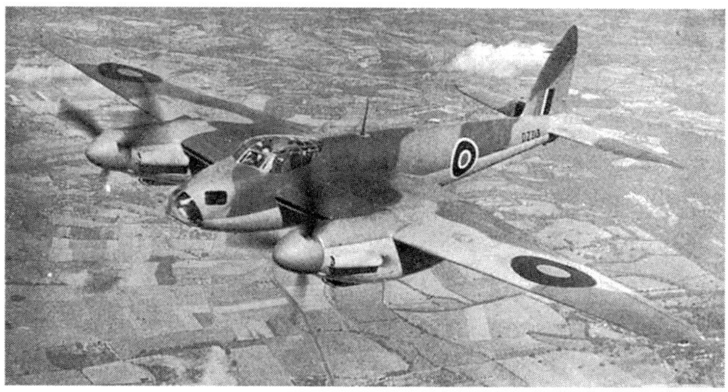

Fig. I.8 British Mosquito.

It was a British multi-role combat aircraft that served during the Second World War and the postwar era. It was nicknamed "The Wooden Wonder." In England, fuselage shells were mainly made by the furniture companies. "The manufacturers had dexterous young women ironing the (unusually thin) strong wood veneer product. Many of the other parts, including wing spars, flaps, flap shrouds, fins, leading-edge assemblies, and bomb doors were produced in High Wycombe, Buckinghamshire in the United Kingdom that was well suited to these tasks because of a well-established furniture manufacturing industry." [I.1]

In the late 70s and 80s, airframe manufacturers began again to use dexterous young women who applied heat and hand compression tools to "iron" the (unusually thin) strong carbon fiber pre-preg[20] material onto tools.

This new material became known as black metal because it usually replaced metal skins on a one-to-one exchange with metal parts. The substructure almost always stayed metallic.[21] As composites advanced, it became apparent that the new material

20. The term pre-preg or prepreg consists of a combination of a matrix (or resin) and fiber reinforcement. It is ready to use in the component manufacturing process.

21. A composite skin with a metallic substructure adds a significant degree of difficulty to the drill portion of the D/C operation. When the drill breaks through the composite material and engages the metallic substructure, the chip spiral exits through the softer composite material. The chip spiral can tear the softer composite material, which damages or opens the hole beyond its allowable control limits.

could be used and manufactured in previously unforeseen ways. Design of bonded unified structure, Z-pinning, bonding of dissimilar materials, embedded antennae and sensors, and other enhanced composite applications for manufacture and flight are being tested. In some cases they are migrating past the test and evaluation stage onto existing production lines and into airframes.

These new applications that will reduce fastener counts and, therefore, D/C has been slow to migrate into production. It will be a long time before the smells, sounds, and sight of drills making holes en masse disappear from the military or the commercial airframe manufacturing and assembly floors. That is all right. I began my career at North American Aviation as an apprentice tool maker in 1965. Over the years, I have come to love the smells, sounds, and sights as evidence of a vibrant living factory.

End Notes

i In fact, rolling or moving assembly lines similar to the NORAIL line are returning. The predominant assembly lines of the last five decades are being replaced for a variety of reasons. One reason is the hundreds of crane moves that are necessary in a boxed assembly line method to move components and subassemblies from assembly fixture to assembly fixture.

ii The Disruptive Technology Office (DTO) is a funding agency within the U.S. intelligence community. It was until recently known as Advanced Research and Development Activity (ARDA). ARDA was created in 1998 after the model of the Defense Advanced Research Projects Agency (DARPA) by the director of central intelligence and the Department of Defense, and took responsibility for funding some of DARPA's projects.

iii It will surprise those unfamiliar with a modern airframe assembly line that there does not seem to be more of a reliance on ongoing quality systems that rely on algorithms and advanced mathematical predictability formulas, which will minimize or eliminate the need for complete line shut downs.

iv ED is one of the most critical components related to hole integrity. Fasteners are subjected to random cyclic loading. In a place or places where there is short ED, fatigue or creep failure occurs. The failure progresses through the constraints and contacts. The cracks can slowly degenerate the structure if they are not detected and fixed. In extreme cases, short ED can result in catastrophic failure of the entire structure.

Reference

I.1 RAF Wyton—De Haviland Mosquito, http://www.raf.mod.uk/rafbramptonwyton/aboutus/dehavilandmosquito.cfm, Retrieved June 6, 2012.

Chapter 1
Historical Perspective: Why Airframes Have Holes

A people without the knowledge of their past history, origin, and culture is like a tree without roots.

—Marcus Garvey

1.1 Introduction

The first airframe did not have fasteners.[1]

Fasteners specifically for airframes were invented and have evolved as new material types have migrated into the airframe. A cycle of increasing material types combined to drive requirements for increasing fastener types. Therefore, to understand why airframes have holes, it is essential to understand the link between airframe material and fasteners.

Figure 1.1 illustrates the continuous design improvement cycle and its effect on materials, tolerances, and fasteners that drives manufacturing complexity and change.

Development of new material types always has enabled the invention or advancement of technology. The history of the modern airframe and its manufacturability is also linked to development of new types of material. The invention and evolution of the airframe fastener and the technology to facilitate its installation is the result of introducing new materials onto the airframe.

1. The first airframe did have screws and nails.

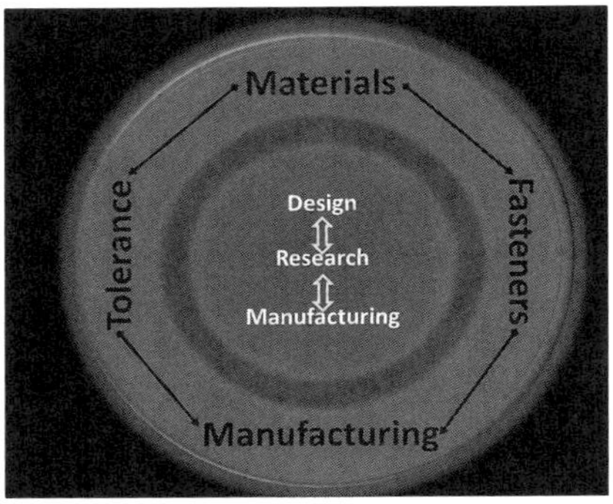

Fig. 1.1 The design improvement cycle and its effect on manufacturing.

Airframes are the most critical part of the airplane. An airframe can perform controlled flight without an engine (gliders). An engine cannot perform controlled flight without elements of the airframe. However, engines and airframes have a symbiotic relationship, and their histories are inextricably linked. More powerful engines are a prime motivator that drives the search for better materials, which drives fastener and manufacturing development.

1.2 The Airframe

Airframe manufacturers have had a love hate relationship with the airframe assembly process since the original airplane flew in 1903. The Wright brothers used cloth over wood and glue to hold the airframe together. Some screws and nails were used, but the strength of the airframe was derived from glue and its ability to hold the cloth skin to wooden skeletal substructures. Orville and Wilbur Wright designed and flew the first fixed-wing design that incorporated the mechanisms necessary for controlled flight. Their original design used their patented wing warping for roll control of the airplane. They also incorporated a canard and rudder. Their bi-wing design, fuselage, canards, rudders, and wing warping combined to be the genesis of modern airframes [1.1].

The evolution continued into World War One when many components formerly outside the fuselage were internalized. The engine was enclosed in the fuselage or in a covering called a cowl. The pilot sat inside a cockpit. Control cables, fuel tanks, and electrical systems began to be attached inside the fuselage.

A new lightweight material called aluminum became commercially available and economically viable. Aluminum became a high-strength alternative to wood and cloth for components of the airframe. Its integration into airframe designs drove new fastener and manufacturing technology.[2]

2. Although this book is about airframes, it should be noted that the engine block of the original Wright Flier that made the historic flight in 1903 was made with a cast aluminum alloy.

1.3 Aluminum

The history of the airframe cannot be written without the inclusion of the material that was the main contributor to the modern airplane's development.

Humans knew for centuries that aluminum is a naturally occurring and abundant metal found most places on earth. Extracting the metal from the earth in an economical way was the challenge. Hans Christian Oersted, a Danish chemist, found a way to produce minute quantities of aluminum in 1843. By 1845, another chemist, Friedreich Wöhler, was able to improve the method to produce larger sample sizes. Henri Étienne Sainte-Claire Deville improved Wöhler's process, which resulted in the commercially and economically viable production of aluminum. In seven years from 1852 to 1859, Deville's production process reduced the price of aluminum from ~ $2640/lb ($1200/kg) to ~ $88/lb ($40/kg). The price of aluminum was headed in the right direction for widespread consideration in product design and manufacture [1.2].

Another chemist, Charles Martin Hall, invented and patented his inexpensive method for aluminum production in 1889, which brought the price and purity into widespread commercial use. By 1914 Charles Martin Hall had brought the cost of aluminum down to 18 cents/pound (~40 cents/kg). The company he founded in 1888 with financier A.E. Hunt is known as the Aluminum Company of America (ALCOA).

The pursuit of aluminum as a commercially and economically viable material resulted in near simultaneous discoveries in Europe and the United States. Inexpensive processes for extracting aluminum were invented by French chemist Paul L. T. Heroult in 1886 and Austrian chemist Karl Joseph Bayer in 1888. The methods that Hall, Heroult, and Bayer developed are still used to produce nearly all of the world's aluminum [1.3].

The application and integration of aluminum into the airframe began before the First World War. Components such as covers, seat frames, cowlings, brackets, and fuselage structural tubing were common by the beginning of the First World War. Most of these components were either welded or bolted as a means of attaching two components together.

During World War 1, the Germans and the Allies began using Duralumin. Duralumin was invented by the German Alfred Wilm in 1903 and developed in the United States. Duralumin's name is a composite of the first word in the name of the factory where Alfred Wilm worked as a metallurgist (Dürener) and the first six letters in aluminum. It was the first heat treatable, high-strength aluminum alloy. Duralumin was first used in airships such as the famous Zeppelin Hindenburg that met its demise in the fiery crash at the Lakehurst Naval Air Station, New Jersey, on May 6, 1937. Figure 1.2 shows the fire-damaged Duralumin cross brace from the Zeppelin airship Hindenburg salvaged from the crash site [1.4].

As late as 1980, all aluminum was still called by its "nickname" Dural. The name stuck, even though the original Duralumin (2017-T4) had been replaced by higher-strength aluminum alloys in both the airframe (7075-T6) and airframe tooling (6061-T6).

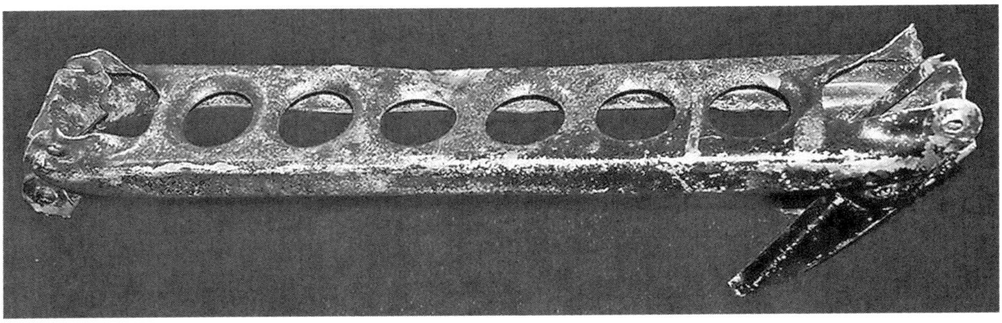

Fig. 1.2 Zeppelin Hindenburg's fire-damaged cross brace.

Although a wartime secret, Duralumin spread throughout the postwar aircraft industry in the 1930s. Duralumin was introduced as corrugated skin in the Junkers J 3 monoplane airframe [1.5].

1.4 Rivets

Enter the rivet. There is archeological evidence that the rivet in primitive form has been around for several millennium. Copper nails were driven through two pieces of material that needed to be joined. The protruding point of the nail that exited the material was bent over or expanded to hold the two pieces tightly together.

When aluminum emerged as the future material for airframes, engineers turned to rivets as the means to hold the structure and skin together. Aluminum rivets are still used today on many aircraft with little or no deviation from the original concept. A hole is drilled slightly larger than the body of the rivet. A rivet is placed through the hole so that it protrudes through the other side of the two pieces of material to be joined. The rivet is expanded to fill up the hole and form a cap that is opposite the head and compresses the two pieces of material together. Figure 1.3 illustrates the two-step process showing a rivet before and after compression.

The use of aluminum rivets to join structure and skin in aluminum airframes created the need for aerospace-specific manufacturing requirements. Design requirements included hole size controls for specific rivets, and edge distance controls for hole placement. In addition, the burr that remained as a residue of the drilling process needed to be removed before insertion of the rivet. Those first aerospace design requirements for airframe assembly that defined the manufacturing controls for holes were the beginning of today's manufacturing specifications for hole quality. Some of the controls established in the 1930s continue without change today.

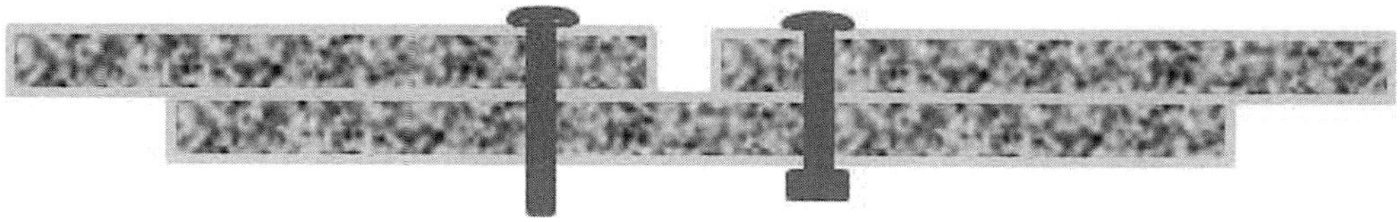

Fig. 1.3 Rivet before and after compression through two pieces of material to be joined.

The next evolution in rivet design and manufacturing complexity came in the form of an invention used by Howard Hughes. The flush head rivet was incorporated on his Hughes H-1 Racer to reduce drag. Earlier rivets had button tops that protruded into the airflow and created drag. This was not a problem because most airplanes prior to 1930 were not going fast enough to have the protruding rivets affect performance—or protruding rivets were not recognized as inhibitors to performance.

Howard Hughes' incorporation of flush head rivets was hardly noticed by anyone but the aerospace community. However, it had a profound effect on the manufacturing component of the aerospace community because it elevated the complexity of the manufacturing process. The incorporation of countersinking as part of the drilling process required acceptable countersink depth definition, tools, and process controls to ensure conformance to design requirements. For flush rivets to be a performance enhancement, the skin has to remain the same thickness that was present for button top rivets. Otherwise, the weight difference from the increased skin thickness would mitigate the performance benefit of the flush fastener. Controls for countersinking depth were needed to prevent intrusion into the substructure of the airframe. Figure 1.4 illustrates the proper depth of a countersink and over-countersunk hole.

Fig. 1.4 Illustration of proper countersink depth and a countersink too deep.

Overcountersinking a hole diminishes the strength of the rivet and the surrounding material to hold the airframe together. Catastrophic failure can occur when an airframe has improperly installed rivets or fasteners as the result of over-countersunk holes or incorrect hole size. The failure of a rivet under the right circumstances can result in a zipper effect. The zipper effect occurs when failure of one rivet propagates the failure along the rivet row to cause failure in every rivet.

The most dramatic result in recent times was Aloha Airlines Flight 243, when in 1988 a Boeing 737 had its roof rupture at 24,000 feet. Corrosion had stressed the rivets beyond their design capacity. When the first small control area broke away, rapid decompression tore away a large section of the aircraft from just behind the cockpit to the forewing section. This failure sequence is called the *zipper effect*. The same zipper effect can be caused by the improper D/C of attach holes in a new airframe. Figure 1.5 shows the damaged upper fuselage of the Aloha Airlines Flight 243 [1.6].

Fig. 1.5 Roof rupture of Aloha Airlines Flight 243.

Countersinking holes had added enhanced performance to the airplane while adding complexity and process-critical aspects to their manufacture.

As performance demands on airframes increased, new types of fasteners replaced rivets and began to result in tighter-tolerance holes as well as more precise placement of holes. The new fasteners were inserted into holes drilled closer to their diameter, which added complexity to the countersink process. When a soft rivet is inserted into an oversized hole, the process of squeezing the rivet allows it to conform and fill the hole and the countersink. A certain amount of variance is "forgiven" because of the soft material, the oversized hole, and the manufacturing process. When a hardened high-performance fastener is inserted into a near net-to-fastener sized hole, the fastener follows the hole. If the centerline of the hole is not 90° to the surface that has been countersunk, two defects occur. The countersink is not flush to the surface. This results in one side of the fastener sticking above the surface while the opposite side is below the surface. In aerospace jargon this is called a "Smiley Face." As cute as the name sounds, it can be a serious defect that weakens the fastener. The second effect of a hole not 90° to the centerline of the surface is a gapped nut on the back of the fastener. This defect further weakens the fastener. All of the advancements in fastener technology when combined with the development of specialized fasteners for specific applications resulted in a complex manufacturing process that increased the cost to manufacture the airframe.

1.5 Assembly Tooling

The evolution of tooling to meet the stringent requirements of precise hole placement, hole diameter, and countersink control evolved through the 1950s. The complex system that enabled the control of the key elements of airframe assembly matured in the 1960s to build the airplanes of the next three decades.

In El Segundo, California on Imperial Highway not far from Los Angeles International airport was North American Aviation's Building 101. During the South Bay area of Los Angeles' Golden Age of aviation, Building 101 was the epicenter for construction of airframe assembly tooling. Commercial and military assembly tooling was built side-by-side for every component for most of the airplanes produced in the United States and Europe. Entering the building in 1965 would reveal assembly tooling for the new generation of commercial airplanes such as the Boeing 727 and the Ling-Temco-Vought A-7 military airplane. The assembly tools built in Building 101 carried the genealogy that would define the manufacturing process for assembly of future aircraft. Most of their defining characteristics were related to key elements of assembly that had a direct relationship to hole placement, hole diameter, and countersink control.

Assembly tools resembling large steel boxes weighed tons. They contained details meant to position precisely and rigidly hold the parts of the airframe in position. They also contained positioning devices that held drill bonnets and drill templates in proper relationship to the parts. The bonnets and templates facilitated the precise drilling of holes by airframe mechanics.

Out on the airframe assembly floor, a new generation of hand tools aided the mechanic. Drill cups helped drill close-tolerance holes 90° to the surface of the airframe in progressive steps. Countersink cages were used to precisely countersink holes normal to the surface of the airframe while controlling depth. Figure 1.6 shows the drill cups and countersink cages still used today.

Fig. 1.6 Countersink cage and drill cup used for precision hand D/C of holes.

All of the assembly tooling was supported by precision measuring devices for construction and calibration. Template makers and model makers made master models and gauges to control and maintain the dimensional integrity of the assembly tools and their locating details. Drill bonnets and templates were constantly in construction or repair to maintain the integrity of the holes they served. The tooling that facilitated the hand assembly of the airframe had reached its highest form of precision. Parts were hand trimmed, positioned, drilled, countersunk, and fastened together in progressively larger steel boxes. Refinements continued with the advent of more precise instruments to perform measurement and calibration of the large assembly tools. However, the basic airframe assembly tooling looked the same when constructed in the 1990s as it did when it was constructed in the 1960s. The hand tools the mechanic used to facilitate quality holes, countersinks, and fastener installation did not change either.

Increased demand for performance drove the inclusion of titanium into airframes in 1960. The advantages of titanium and its growing use on airframes did not change the basic tooling concept for assembly. However, it changed the approach to drilling holes and incorporated pneumatically driven and controlled drilling machines.

1.6 Titanium

M.H. Klaproth, a German chemist, discovered and named titanium in 1793.[3] Klaproth named the new material titanium after the Titans of Greek mythology; it means, "the incarnation of natural strength." His name for the material stuck. However, the element was not successfully isolated until 1910 [1.7].

William Kroll invented the process to produce useable metallic titanium in 1932. His discovery is still called the Kroll process. It is the fourth most abundant material in the earth's crust at 0.62%. Like aluminum it took a metallurgist to develop a process for its extraction. It took nearly a decade for Kroll's laboratory process to be refined for commercial introduction of the first products in the 1940s.

The years from 1938 to 1947 saw a large investment in development of titanium by the United States Defense Department. The attributes of titanium that attracted the investment were the material's high tensile strength to density ratio, high corrosion resistance, crack resistance, ability to withstand moderately high temperatures, and low coefficient of thermal expansion (CTE). Much of the development dollars went to the United States Bureau of Mines where they made various modifications to Kroll's process. DuPont opened the first titanium manufacturing operation in 1948. DuPont's large-scale manufacturing of titanium saw the migration of titanium into the aerospace industry in the 1950s. The aerospace demand for titanium has fueled its growth. Aerospace continues to be the primary user of titanium, consuming 80% of production.

The first aircraft to make extensive use of titanium in its airframe was the Lockheed Skunkwork's SR-71 Blackbird. Lessons learned from the Blackbird paved the way for

3. William Gregor, a British chemist, also discovered titanium independently in 1791 and named it gregorite after himself.

its extensive use in today's military and commercial aircraft. The Airbus A380 uses more than 146 metric tons of titanium, with 120 tons used in the airframe.

As the use of titanium increased, the difficulties for manufacturing assembly increased. Titanium introduced difficulty into the manufacturing process, specifically with regard to the drilling of holes.

1.7 Hand Tools and Drills

The tooling philosophy for the control of hole position and diameter would not change with the introduction of titanium and other high-strength materials such as Hi Tuff steel.[4]

The new materials were a boon to the strength and performance of the aircraft but drove requirements for new types of drills, drill motors, and processes to successfully drill holes for fasteners. The new material's toughness could not be hand drilled through the thicker sections of the airframe because of the speed of the motors and pressure required to feed the drill. The drills used at the time of the tougher material's introduction were not adequate to remove the tougher material. Therefore, a new type of pneumatically powered and fed drill motor incorporating a new process was developed. The new motors required stronger drill bonnets to hold them in place due to their weight. The added pressure was facilitated by an increase on the shop floor of larger ¾-in (19-mm) air supply lines. The supply lines were mostly ½-in (12.7-mm) before the introduction of the new materials because the air supply through the ½-in (12.7-mm) line was sufficient to push a drill by hand through aluminum. The new air lines, drill motors, and drill types to drill holes in the tougher titanium and other materials broadened the supply base. It increased the need for larger air lines, compressors, and receiver tanks to store compressed air. It also complicated the manufacturing process and added internal support departments for the new drills, motors, facilities maintenance, and support. The added complexity, supply chain, facilities upgrades, and complexity added cost to the airframe to enhance airplane performance.

New processes to drill through the new material had to be developed, defined, and transitioned to the shop floor. One of those processes was peck drilling. Peck drilling is a process to drill a hole in stages. The drill motor is installed into the drill bonnet. When the drill motor is turned on, it begins spinning at the motor's set revolutions per minute (rpm) and extends rapidly to a distance just short of the material. The feed rate slows to the specified feed rate for the drill size and type. The drill tip engages the material at the preset speed and feeds to a specified distance below the surface. The drill motor then retracts rapidly to withdraw the drill from the material to clear the chip. The drill tip and the hole are flooded with coolant while the drill is momentarily withdrawn from the hole. The first drill depth breaks the skin of the material. After extraction, the drill rapidly extends back into the hole just short of the first drill depth and then reduces the feed rate back to

4. Lescalloy Hy-Tuf steel is a high-strength steel used in landing gear, flap tracks, and structural components such as the T-38 Talon's landing gear wing rib.

the preset for the type material and drill diameter. The cycle is repeated until the drill exits the material and a uniform hole is produced.

Pecking is required in materials such as titanium due to its toughness. Precise hole diameters that are consistent all the way through the hole are produced by pecking. Titanium materials also cause chip loading of the drill that leads to a broken drill in the hole if pecking is not used. Breaking a drill in titanium material causes extraction difficulties that can damage the hole and in extreme cases damage the part beyond repair.

An example of the difficulties inherent in drilling titanium and the differences between drilling titanium and aluminum is the first F/A-18 C/D Swiss airframe. In the mid-1990s the first Swiss F/A-18 C/D airframe came down an assembly line that had matured and produced over 500 units of the venerable airframe for the U.S. Navy. The Swiss airframe had one difference. Instead of cast and machined aluminum bulkheads, they had cast and machined titanium bulkheads. Mechanics and masters trained at drilling through aluminum or composite skins into aluminum bulkheads used the same process and tools on the titanium substructure. The result was the complete replacement or repair of most of the bulkheads and other titanium substructure. The mistake was costly and time consuming.

The added burden of maintaining the more complex pneumatic tools for drilling titanium and other tougher materials increased manufacturing infrastructure. The more complex motors needed maintenance, calibration, and drill set up control that added to the complexity and cost of airframes.

A broader array of drills to meet the demands to D/C titanium meant increased inventory, distribution, quality control, and process controls that also added cost to airframe production.

Using pneumatics for drilling precision holes through tough materials was beginning to show its limitations. The precision demanded by higher airframe performance was stretching the limits of pneumatic drill motors. The physics of blowing air over vanes to initiate and maintain the proper speeds specified for drilling precision holes is affected by a number of factors. In aerospace, as more precise holes became the norm, pneumatic drills were required to be calibrated. The speed had to be calibrated and certified for a speed range. Different pneumatic drills were specified for drilling different hole sizes. For example, higher-speed drills were specified for small holes in aluminum and lower-speed ones for larger holes or holes in tough materials such as titanium. The calibrations and certifications were done in the calibration laboratories with a constant airflow at 80 pounds per square inch (psi), without being under load.

When the drill motors (pneumatic drills) went back to the shop floor, the shop air supplied to the motors varied from as low as from 40<60 psi up to as much as 100 psi. The air supply was constantly varying as air motors were started and stopped. Therefore, the specified speed rates were constantly in flux. Compounding the variability of speeds was the pressure applied to the drill motor by the airframe mechanic. Specified feed rates were varied by the pressure the mechanic placed on the drill. Therefore, part of the equation for a good hole in any material was the art of the mechanic and his or her ability to "feel" the cut of the material. Tougher materials intensified the reliance on the mechanic's sensitivity to material removal. Good mechanics made good holes most of the time.

The introduction of titanium as a substructure material introduced another challenge to the mechanic and the quality of the airframe. Drill starts are a defect that occurs in military airframe assembly, although it also exists to a lesser degree in commercial airframe production.

Military airframes are assembled from the inside out. The skins, doors, and covers of the airframe close out the interior after all the wires, tubing, electronic gear, and other interior systems and components have been installed. This is called the "stuffing" of the airframe. Some of the stuffing is mounted onto or goes through the bulkheads and other substructure details. The skins are mounted to the substructure and drilled. The titanium substructure requires a greater force to drill and break through to the other side. The force applied to the tip of the drill by the mechanic, when combined with the hard chip created by the flutes of the drill bit, causes the drill tip to uncontrollably and rapidly exit the hole. If there is a piece of stuffing in the Line of Sight (LOS) of the drill and close enough to the exit point of the hole, the drill tip strikes the stuffing, causing a drill start, which is a defect. Defects of this nature are serious. They result in evaluation of the defect to determine the impact the drill start has to the strength and performance of the stuffing component the drill bit tip has struck.

Drill starts are not a new defect; they occurred when softer materials were drilled. Tools were invented that slid onto the drill and were used to limit the breakout extension of the drill tip. Figure 1.7 shows a common drill stop tool for preventing drill starts.

Fig. 1.7 Drill stops of varying lengths with attach screw.

Drill stops worked for softer materials such as aluminum. However, the tougher chip of titanium could still pull the drill through the exit point of the hole and occasionally defeat the strength of the drill stops.

As drill motors encountered tougher materials, additional pressure applied also caused physical manifestations. As pressure increased to push the drill through the material, the result over time has been a rise in injuries to mechanics from pressure to the wrist. Even though the thicker materials were drilled by mechanically fed, pneumatically powered peck motors, the mechanic still drilled and countersunk the majority of holes.

About the time titanium and the manufacturing process for its successful inclusion in airframe assembly had matured; a new material began to emerge. Around the beginning of 1980, carbon fiber (composite materials) began to migrate onto the airframe.

1.8 Composites

Outside the South bay door of Northrop Grumman Corporation's composite manufacturing center of excellence in El Segundo, California is a small plaque. It identifies the building as the location for manufacture of the first composite part that was installed on an operational aircraft. The aircraft was the Northrop F-5.

In the beginning, composites were called black metal because composite components replaced metal components on a one for one basis. Its properties were highly desirable, especially for skins where its weight and strength were significantly better than the aluminum skins it replaced.

Commercial and military air vehicles, such as airplanes, helicopters, unmanned aerial vehicles (UAVs), missiles, drones, and rockets placed a premium on high-strength, lightweight materials. Metallic components for air vehicles had satisfactory mechanical properties. However, the weight of metallic components limited the aggregate lift capacity of the vehicle. The higher the weight of the air vehicle or rocket, the less fuel, cargo, and ordinance it could carry.

The first research projects to investigate the potential for building airplane parts from composites was initiated at Wright-Patterson Air Force Base in 1943. The research and development effort resulted in the first Glass Fiber–Reinforced Polymer (GFRP) fuselage. The airplane that incorporated the GFRP fuselage flew in 1944 [1.8].

Fabrication processes and material advances were developed for GFRP parts that sped the design and distribution of composite parts for airframes. Two advances were the development of pre-impregnated sheets of glass fibers in a partially cured resin, or pre-pregs, and the development of tooling processes for the curing, cutting, and trimming of composites. These advances simplified the GFRP component fabrication process for manufacturability.

The advantages of GFRP technology drove advancements and refinements in the technology. By 1958 the French were producing helicopter blades for the Alouette II.F using the new composite material. Approximately 90% of the commercial market for composites is still comprised of GFRP.

Another advantage of composite material over other materials is its design flexibility. The stacking of multiple layers of cloth cut in different shapes tailored the thickness of the parts. The raw material was malleable and could be shaped easily before cure. More-difficult shapes could be attained using composite materials.

The development of composite materials coincided with the development of NC technology for machine tools. Before NC machines, airframe shapes were attained through a complex process using loft books, loft lines, and hand-filed templates positioned at intervals. The templates were mounted on plates, filled with plaster, and swept into the filed templates' loft lines by hand. Casts were taken from the finished models, and tools for airframe parts were produced from the models. The new NC technology used a computer to drive machines. The spindle of the machine contained a cutter that could precisely cut the fabrication tools for composite parts directly from NC programs linked to the engineering design. The synergistic effect of these two technologies elevated the capability of both the design engineer to create, and the fabrication shop to build, precise and complex shapes.

The composites manufacturing process was laborious, and the new material was more expensive than metal. However, beginning around 1980, weight and performance characteristics drove increased incorporation of composites onto airframes during the next three decades.

The Beechcraft Starship designed by Burt Rutan in the late 1970s was the first certified all-composite business aircraft. Today's commercial aircraft such as the Boeing 787 and military aircraft such as the F35 use composites extensively in their advanced airframes. Designers are also beginning to incorporate composites into high stress areas formerly reserved for metallic materials. The Ares I-X space launch vehicle that was successfully launched on October 28, 2009 incorporated a composite interstage section. Figure 1.8 shows the 2009 launch of the Ares I-X with its composite interstage during earth departure.

Fig. 1.8 Ares I-X launch with designation of composite interstage.

The design and test of an all Composite Crew Module (CCM) and the design and launch of the composite Max Launch Abort System (MLAS) in July 2009 are recent examples of the migration of composites across a wide range of air and space vehicles. Figure 1.9 shows the all Composite Crew Module (CCM) and launch of MLAS from Wallops Island, Virginia.

Fig. 1.9 CCM and MLAS launch.

1.9 Tooling and Composites

The introduction of composites challenged the manufacturing processes developed, designed, and matured for assembling, drilling, countersinking, and fastening metal components.

Several aspects of composite parts challenge the assembly process. They are thickness variability, the untooled rough surface of composites, dissimilarity of material type against which composite parts are assembled, a layer of shim material between the skin and substructure, and supply chain disruption.

The variability of composite parts is derived from two primary sources. Their thickness varies by design and through the variability of the raw material. The designer uses the capability of the composite material to provide variable thicknesses to maximize strength and minimize weight. Thickness by design is derived by adding, reducing, or eliminating plies where needed.

The raw pre-preg material arrives at the fabrication factory to be cut into kits that then are laid up for cure. When the materials arrive, a number of factors contribute to the thickness variability of each cut ply. The variability of the raw material translates into ±8% cured ply variability (CPV) of the finished part. This has become such an important issue that the Air Force Research Laboratory (AFRL), Materials and Manufacturing Directorate, Manufacturing Technology Division (AFRL/RXM) identified CPV as a major negative effect on the assembly of the F-35 airframe. Solicitation Number: NOCA-08-13-PKM states: "The F-35 has stringent requirements to maintain Outer Mold Line (OML) flushness. In doing so, the program incurs significant manufacturing costs and long manufacturing span times. Reducing the Cured Ply Thickness (CPT) variation is an identified area with potential to reduce these manufacturing burdens. A leading contributor to CPT variation is the raw material, prepreg inconsistency."

The variability of each piece of raw material induces variability into the finished part. The greater number of plies (sheets of composite) needed to fabricate a finished airframe part, the greater total variation from the specified design thickness.

The effect of thickness variation when drilling through a composite skin where most of the airframe holes reside is significant. Common drill speeds for drilling holes in composite and aluminum are 4000 to 6000 rpm. For titanium, drill speeds are 300 to 1000 rpm. The speed depends on the size of the hole. Therefore, a small hole going through an aluminum (Al) over titanium (Ti) stack would be drilled at 4000 rpm in the Al and 300 rpm in the Ti. The problem with variability of the composites is derived from the inability of the mechanic to determine when the drill will hit the Ti. If the drill tip encounters the Ti at 4000 rpm, the drill can bind and snap, work harden the Ti, or dull the drill, making it impossible to remove additional metal. This results in a drill change and possible defect.

When composite parts are fabricated, they have a smooth surface and a rough surface. The fabrication process places the material down against the smooth surface of the tool. The rough side is the side away from the tool surface where the vacuum bag is affixed. The smooth surface is almost always the Outer Mold Line (OML), or air flow surface of an airframe.[5]

The rough surface is the surface that is against the substructure. The substructure is a smooth-machined and dimensionally controlled surface. Therefore, the assembly of the airframe requires the addition of a material that can fill the gaps created between the rough surface of the composite and the mating metal surface of the substructure when they come together as an assembly. The material is a liquid shim that goes on wet when the parts are apart; the parts are then squeezed together to mate the skin to the substructure. Various temporary fastening devices are used to apply pressure and compress the skin against the substructure to squeeze out excess material before cure. The shim cures, filling the gaps and forming a tight fit between the two mating components. The liquid shim process is also used when a tooled surface of a composite substructure such as ribs and spars mates against the untooled undersurface of the mating skin. Cured liquid shim is brittle and adds a third layer of complexity to the drilling process.

When a hard Ti or steel substructure, liquid shim, and a composite outer skin are stacked together, hole creation becomes a complex process that challenges the hand-built method of D/C. Add the variability of the composite parts and it becomes even more difficult.

The drill type must default to the type of drill for titanium, which is not optimal for composite. That is because the poly crystalline diamond (PCD) drills and other drills optimized for drilling composites cannot drill Ti.

---

5. An exception exists in many composite launch vehicle components where the power of the engines and rapid ascent mitigates the need for the OML airflow surface rule. Massive power overcomes minor drag concerns.

The second effect of the stacked materials is chip removal. As the drill begins to penetrate the Ti, a chip is formed that exits the hole and can tear the less resilient composite material. Drill steps were added to incrementally increase the drill size until the final hole size was reached. The step drilling process produced an acceptable hole but added cost and time to the assembly of the airframe.

In addition, the brittle liquid shim can crack and chip under the strain of drilling the Ti composite material stack. If it does, the Ti exit chip can become trapped beneath the substructure and skin. The entrapment of chips can damage the liquid shim and the subsurface of the skin. Chip entrapment can also force the skin away from the substructure and cause a defective elongated hole.

Composite parts do not lend themselves to many lean principles. Lean principals such as on-demand part delivery and minimum inventory are the antithesis of the composite part fabrication process. In the 1980s, one minute die changes were the goal to facilitate minimization of inventory. Quick change of dies or tools allowed a ship-set of parts to be delivered to the shop floor in the sequence they were needed. Metal parts could be made in smaller lots because the tools for machines, and the dies that made the parts to be assembled, could be economically changed rapidly to feed the assembly line as they were needed.

Composite part production is slaved to several fabrication processes. The largest inhibitor for lean implementation for composite parts is the autoclave. Autoclaves are necessary to cure prepreg composite parts. Autoclaves are extremely expensive to operate. Additionally, their fixed cost to operate remains the same whether one part is cured or a full load is cured. They are large and dominate the shop floor where they are installed. The one ship-set rule for sequential part fabrication is defeated by the autoclave due to the absolute operating rules under which the autoclave operates.

The hand airframe assembly process is at its best when it is in a rhythm uninterrupted by change or variance. Composite and Ti parts and their proliferation as the materials of choice by aircraft designers have stretched the hand manufacturing process to its limit.

As aerospace airframe assembly entered the third millennium Anno Domini (AD), it was apparent that the hand method could not meet the demands of the new airframe designs. Nor could it be flexible enough to respond to the modern demand for multiple airplane types integrated into a single assembly line. The methods, tools, and processes developed during the previous decades would be inadequate to meet the requirements. Something had to change.

1.10 On-Assembly Drilling

On-assembly drilling had been tried before. In 2001, North of Northrop Grumman's St. Augustine facility is a warehouse containing the rusting remains of an on-assembly drilling machine. It was built in the 1980s to drill holes in large sections of the E-2C Hawkeye fuselage. The machine had been mounted to the assembly tool of the Hawkeye fuselage sections but failed to meet the specified requirements for hole position and repeatability. It was abandoned and relegated to the scrap yard.

By the mid-1990s, machine tools and their controllers had reached a point where the application of on-assembly drilling could meet the requirements of the modern airframe for hole position and diameter quality.

On-assembly drilling offered the answer to the struggles and challenges that had stretched the hand methods to D/C holes. The opportunity for the first trial and application came on the vertical stabilizer for the F/A-18 E/F. The new assembly line was in construction. The elements of a new technology were present to meet the demands of composite Ti stacks of material. The first machine to combine the technology elements into a machine that drilled holes on-assembly in a military airframe occurred in October 1997 on an F/A-18 E/F vertical stabilizer. The machine was called an Electronic Gantry Applied Drilling System (EGADS).[6]

EGADS had an NC controller that was programmed to drill holes from data directly derived from the engineering data and translated into position and drill commands. EGADS incorporated a vision system, integrated coolant, air supply, and a pressure foot to maintain compression between skin and substructure. The Human Machine Interface (HMI) included the first interactive graphics for operator instruction, alert notices, and email alerts for calibration. Figure 1.10 shows the EGADS—the first on-assembly automated drilling machine for military airframes.

Fig. 1.10 The EGADS on-assembly drilling machine.

6. U.S. patent numbers 6,033,226; 6,231,280; 6,232,736; 6, 292,712.

The in-one-pass without the need for step drilling improved hole quality and throughput. Countersinking was added to the operation six months later and resulted in a high-quality, full-size drill and countersunk hole in one pass.

Other EGADS followed, with the most notable designed by Northrop Grumman engineer Frank Chang. The design incorporated the best features of the fixed EGADS with mobility. The machine dubbed the Mobile Gantry Applied Drilling System (MOGADS)[7] could be moved from fixture to fixture by means of a pneumatic lift device. It precisely locked onto the fixture and aligned itself with the fixture reference system automatically before D/C.

Just before the year 2000, all of the advances in machine technology combined to build an on-assembly drill countersink machine called the JSF Gantry Applied Drilling System (JGADS).

On March 1, 1999 Lockheed Martin issued the following press release.

> *"The Lockheed Martin Joint Strike Fighter (JSF) team has developed an advanced drilling system that significantly reduces manufacturing time and costs in a lean production environment.*
>
> *The JSF Gantry Applied Drilling System (JGADS) is part of Lockheed Martin's Airframe Affordability Demonstration (AAD) program. AAD is validating the producibility of the team's JSF aircraft and the savings associated with new equipment and processes.*
>
> *"JGADS has evolved from previous concepts used by JSF team members and is more advanced than anything else that exists today in the aerospace industry," said Frank J. Cappuccio, vice president and program manager of the Lockheed Martin JSF team.*
>
> *This system has the precision of large, heavy gantry drilling machines but with the portability, flexibility and lower cost of lightweight robotics. It drills and countersinks fastener holes precisely, using a laser-positioning system, magnetic motors and a patented "pressure foot" capability.*
>
> *This kind of accuracy speeds up assembly and creates close-fitting panels and fasteners, resulting in smoother surfaces and smaller gaps. These surface features not only reduce drag but also are key in meeting JSF survivability requirements.*
>
> *In initial use, JGADS registered only three defects per million, surpassing a world-class drilling quality standard of 3.4 defects per million; this translates into significant savings in scrap and rework. The results were so encouraging that the F-16, F-22, F-2 and T-50 (KTX-2) programs are evaluating JGADS-type systems for their assembly operations.*
>
> *"Even though JGADS handles large parts, it can be easily disassembled, moved and reassembled. The system's versatility and portability allow it to be placed directly in component assembly areas on the production line, eliminating the need for dedicated machining centers and extra parts movement.*

7. U.S. patent number 5,836,068.

JGADS uses commercial off-the-shelf, user-friendly software that can be programmed directly from the computer design model in one-tenth the time required for conventional robots. Using virtual production modeling, the entire operation is validated by computer, eliminating the need for checkouts or "dry runs" at any time.

It takes 10 minutes to set up component structures in the work station—a 90-percent time savings compared to conventional systems. With JGADS, the JSF team expects a 50-percent reduction in direct man-hours versus use of manual tools. Additionally, the team estimates a drilling tooling reduction of 75 percent through elimination of tool templates, drill plates and portable power tools. The system is easy to maintain and requires no special foundation, further reducing tooling costs." [1.9]

Figure 1.11 shows the author operating the JGADS automated drilling machine.

One of the breakthrough features of this machine was the invention of a software system that could use CATIA (Computer-Aided Three-Dimensional Interactive Application) engineering models and translate them within minutes to a Microsoft Excel program. The Microsoft Excel program operated the machine to perform positioning and drill countersinking operations without the aid of an NC program or post process.

The age of the exclusively hand assembled airframe was over. The age of on-assembly D/C machines integrated into the assembly process had arrived.

Fig. 1.11 JGADS.

References

1.1 Wright, Wilbur, Orville Wright, Russell Freedman, *The Wright Brothers: How They Invented the Airplane*, Smithsonian, March 1994.

1.2 http://www.historyofaluminum.com/, Retrieved May 2, 2012.

1.3 http://sam.davyson.com/as/physics/aluminium/siteus/history.html, Retrieved May 6, 2012.

1.4 Feigenbaum, Aaron, *The Hindenburg Disaster*, Bearport Publishing Company, Inc., New York, NY, 2007.

1.5 Kay, Antony L., "Junkers Aircraft and Engines 1913–1945," http://tocs.ulb.tu-darmstadt.de/125360509.pdf, Retrieved May 2012.

1.6 National Transportation Safety Board, Aircraft Accident Report, Aloha Airlines, Flight 243, Boeing 737-200, N73711near Maui, Hawaii, April 28, 1988, NTSB Number: AAR-89-03, NTIS Number: PB89-910404.

1.7 Housley, Kathleen L., *Black Sand, The History of Titanium*, Metal Management Aerospace, January 2007.

1.8 Vinson, Jack R., *Sandwich structures 7: advancing with sandwich structures and materials*, "Sandwich structures past, present, and future," Part 1, 2005, pp. 3–12.

1.9 "Advanced Drilling System Supports Lockheed Martin JSF Program Affordability," Fort Worth, Texas, March 1, http://www.prnewswire.com/news-releases/advanced-drilling-system-supports-lockheed-martin-jsf-program-affordability-72845517.html. Retrieved May 2, 2012.

Chapter 2
Airframe Manufacturing Cost Drivers

Work expands so as to fill the time available for its completion.

—C. Northcote Parkinson

2.1 Introduction

The direct cost of an airframe is a multiple of the hours of touch labor and material cost used to complete the fabrication and assembly operations (including paint). Part fabrication supplies the individual components and work pieces to the assembly floor for joining into a completed airframe. Therefore, the majority of the direct cost for fabricating the pieces, parts, and components supplied to the assembly floor resides in the material. Since the parts, pieces, and components already contain most of the material and labor cost for their fabrication, the primary cost driver for the assembly of an airframe is the touch labor used to position, assemble, and fasten together the supplied parts, pieces, and components.

Added to the direct cost of the airframe are the overhead costs to operate the factory and assembly tooling costs necessary to produce and maintain the production effort.

The cost of assembling a military airframe is approximately 65% of the cost of the airframe with part, piece, and component fabrication accounting for approximately 35%. Therefore, touch labor hours in assembly are the largest contributor to the cost of modern airframe production.

Within the assembly process there are many touch labor operations. They are categorized as part placement, stuffing (wires harnesses, tubing, internal components, etc.), drilling, countersinking, fastener installation, and paint. The largest contributors to direct labor cost are the drill, countersink, and fastener installation operations. These three categories account for approximately 65% of the cost of assembly.

Added to the direct cost of labor for drilling, countersinking, and fastener installation are the indirect costs attributable to these operations such as their contribution to lost time injury cost (80%) and cost of quality (80%) that adds burden to the hourly wages, benefits, and operating costs of the assembly process.

Additional cost burdens to the touch labor for hand processes involving drilling, countersinking, and fastener installation are the creation and maintenance of standard tooling, special tooling and support equipment, assembly tooling and drill tools, training, and maintenance.

Very little has been written with regard to these breakout numbers because each airframe producer considers their exact numbers and breakdowns proprietary and competition sensitive. However, the universal understanding that the primary cost driver for the assembly of airframes (both military and commercial) is rooted in drilling, countersinking, and installation of fasteners has driven the development and installation of automated systems to reduce touch labor costs. All the major, and to some extent the smaller, airframe assemblers recognize the contribution drilling, countersinking, and fastener installation make to the cost of the airframe (and to the costs associated with the ancillary operations, cost of quality, support, and lost time injuries). The application of D/C machines on assembly to increase quality, consistency, cost, and efficiency has grown across both the commercial and military airframe assembly manufacturers as a means to address these cost drivers and improve the assembly process through consistent parts that fit at assembly.

This chapter will address the various cost effects of the hand drilling operation and its contribution to the cost of assembly of an airframe, and by extension the cost of the airplane.

The hidden costs of assembly that add to the indirect and direct labor cost will be detailed as well as how automated systems eliminate the cost drivers associated with the hand drill, countersink, and fastener installation operations.

2.2 Why Reduce Cost?

Aerospace manufacturers today are searching for technologies and process improvements to gain a sustainable, competitive advantage in the global marketplace. Customer requirements emphasizing greater affordability cannot affect leading-edge technology that enables increased performance attributes such as fuel efficiency and stealth capabilities. The amount of benefits that can be realized by introducing technological improvements into a new [2.1] or existing program have been affected by economic factors over which the industry, airlines, and military have no control. The cancelation of further orders for the F-22 Raptor in recent times and the reduction of orders for the B-2 in the 1980s are examples of the effects of advanced technology outpacing the willingness of the customer to pay the price. Accordingly, the Defense Manufacturing Council (DMC) established several goals to try and control cost [2.2]. The goals include having cost as a design variable on par with traditional performance variables. Designing with an emphasis on economic competitiveness and cost effectiveness became coequal with concerns for production, finance, operations, and support. The Affordable Systems Optimization Program (ASOP) and Affordable Design and Manufacturing (ADAM) program partially funded the development of cost models in the late 1990s that are representative and supportive of design-to-cost manufacturing activities [2.3, 2.4].

The effects of these programs and initiatives rely on compromises to the design of the airframe in deference to cost. In defense acquisitions, performance is still king. And in the commercial, business, and private airplane business, fuel efficiency still rules. Compromises in design to facilitate cost reduction have not realized their full potential as composite materials, advances in computer-aided design software, embedded sensors, and demand for performance enhancements have made each generation of airframe progressively more expensive to produce. The variability of designs customized to meet the specific needs of each customer has further complicated manufacturing requirements where the predominant method of manufacture is human effort. Therefore, the increased complexity required to meet increased customer performance demands has directly affected the labor hours and, therefore, the cost to produce an airframe.

This has not reduced the demand by customers, especially in the global economic climate of 2012, for price reduction as a further incentive to buy more airplanes.

Several contravening forces are acting on manufacturers with the desire to satisfy their customer's demand to lower the price of their product. Complicating the enhanced performance with a lower acquisition price formula is the customer desire to lower the overall life cycle cost of the airplane [2.5].

Therefore, this more-for-less enigma affects the revenue (sales) of the company or its profitability unless it can reduce manufacturing cost. Where cost is the amount of expenditures needed to manufacture the airplane and price [2.6] is the amount paid for the airplane by the customer, profit is the difference between price and cost. Any reduction in price without a commensurate reduction in cost negatively affects the profitability of the corporation. Revenue related to increased sales (resulting from reduced price) is offset by a reduction of per-unit profitability until the lower price-point does not substantiate the increase in sales. Figure 2.1 shows the equilibrium model where increased sales as a result of price reduction begin to negatively affect profitability, assuming cost does not change.

In addition, lowering price to sell more airplanes reduces profit margin while increasing product liability. The more airplanes sold, the greater the liability. The greater liability increases risk to the airplane producer while reducing profit. The liability effects per unit of sales provide further disincentives for the airplane producer to lower cost to increase unit sales without a corresponding reduction in cost.

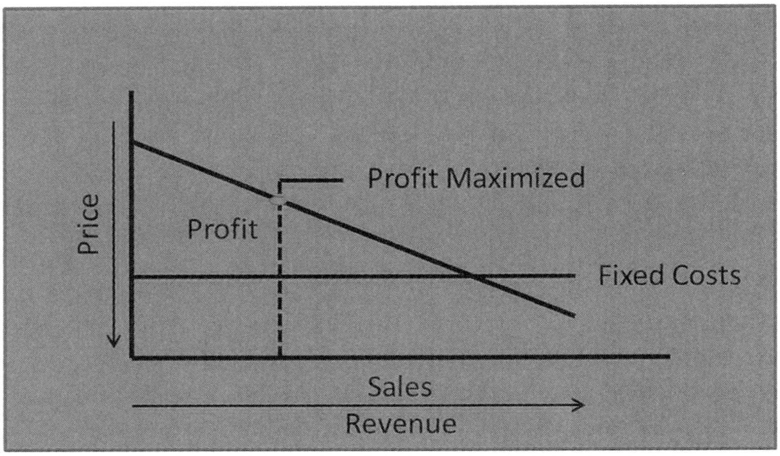

Fig. 2.1 Equilibrium model where increased sales as a result of price reduction begin to negatively affect profitability.

Therefore, the motive to reduce cost is rooted in the airplane producer's desire to increase revenue (price × units) while simultaneously maintaining or increasing profitability (revenue – cost).

2.2.1 The Value Price Profitability Index Commercial

The Value Price Profitability Index (VPPI) for commercial airplanes involves the interrelationship between the airplane producer, airplane buyer, and the customers of the airplane buyer. The interaction between these parties determines the price point that affects the profitability margin of the manufacturer. This interaction is illustrated in Fig. 2.2. The illustration provides the commercial aircraft cost-circular-effect perspective.

Military and commercial aircraft differ in some of the details, but they both are motivated by the desire to increase profitability through increased revenue (sales). And airplane sales are highly price sensitive. Assuming there is demand (price sensitivity), a reduction in price attributable to cost reduction (without performance penalty) would result in increased sales. Even if sales remained flat, reducing cost would increase the profitability of the airplane producer. In order to determine where cost can be reduced without a performance penalty, cost contributors have to be defined and quantified as to their contribution to the airplane's price.

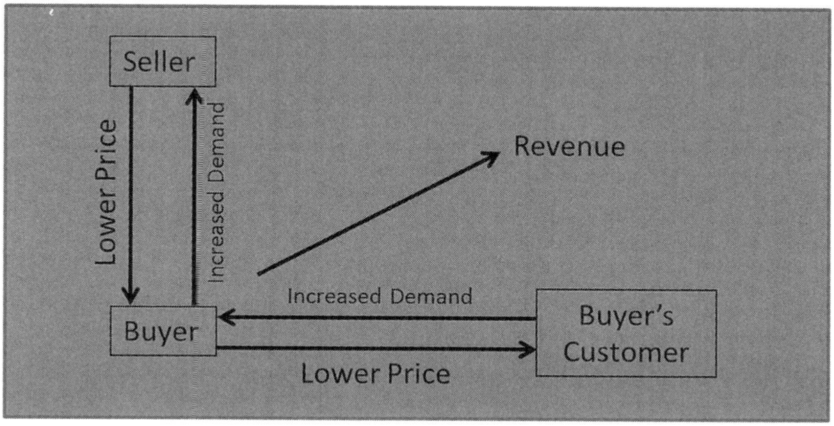

Fig. 2.2 Price point relationship between airplane buyer, buyer's customer, and manufacturer.

2.3 Defining Cost

Identification of contributing cost drivers has to happen before cost reduction can occur. The life cycle cost for an airplane is derived from the six phases of an airplane program:

1. Planning and Conceptual Design
2. Preliminary Design and System Integration
3. Detail Design and Development
4. Manufacturing and Acquisition
5. Operation and Support
6. Disposal

Preliminary cost definitions are derived from the six phases of an airplane program:

CRDTE: Research, Development, Test, and Evaluation
CACQ: Acquisition Cost (CMAN + CPRO)
CMAN: Manufacturing Cost
CPRO: Manufacturer's Profit
COPS: Operating Cost
CDISP: Disposal Cost

Added to the cost as a seventh element of Life Cycle Cost (LCC) is the cost to finance (CFIN) the money used to purchase the airplane. Finance cost is a direct result of price, which is derived from the manufacturer's cost to produce the airplane. CFIN interest on debt attributable to buying an airplane can exceed 15%.

Life cycle cost is the sum of all the six phases of an aircraft program (LCC = CRDTE + CACQ + COPS + CDISP + CFIN).

To determine which elements within the LCC are attributable to airplane production, the total cost to operate the airplane has to be defined. Commercial airplane users and producers often use operating cost, made up of direct (DOC) and indirect (IOC) pieces, to determine the total operating cost of the airplane (IOC + DOC = TOC). Indirect operating costs (IOC) are those costs attributable to airplane, passenger, and cargo related activities and acquisitions. They account for 41% < 44% of the total cost of the airplane. Direct operating costs (DOC) are costs attributable to maintenance, fuel, interest, insurance, and depreciation. They total ~56% of the total operating cost (TOC).

It is important to remember that LCC, IOC, and DOC composites are high-level numbers derived from detailed analysis of finite cost contributors. There are many studies that use extremely complicated formulas and cost contribution analysis methods to precisely determine the content values of LCC, IOC, and DOC. If the reader's curiosity drives them to look at the details that drive the LCC, IOC, and DOC values, a host of information and data is available on the internet. Some of the analysis and data reports are identified in the reference section of this book.

Several cost models of various fidelity levels exist [2.7] as academic, commercial, or industrial products.

The purpose of this exercise is to first determine the prime cost contributors for an airplane and then discover if there are primary cost drivers that can be reduced or mitigated without penalizing performance. Therefore, the higher-level output of all the studies and formulas that collectively arrive at a comparable number are sufficient to discover the focal point to determine the cost reduction targets.

All of the models studied for this chapter conclude that the Direct Operating Cost (DOC) contributes ≈65% to the Total Operating Cost (TOC) of the airplane. An airplane producer can affect the cost of DOC by reduction in fuel cost (efficiency) and manufacturing cost of the airplane.

The overall effect of a manufacturing efficiency improvement effort to reduce the DOC contribution to the Total Operating Cost might seem small. But the airline operating margins are small enough to magnify the Return of Investment (ROI) derived from small price reductions. In addition, if the cost reduction is coupled with improved quality, then the DOC is improved through reduced maintenance and the LCC is reduced through the longer service life of the airplane. Figure 2.3 shows the relationship between price, TOC, and ROI. It is the Airline ROI and TOC (dashed line) vs. Average Yield and Price.

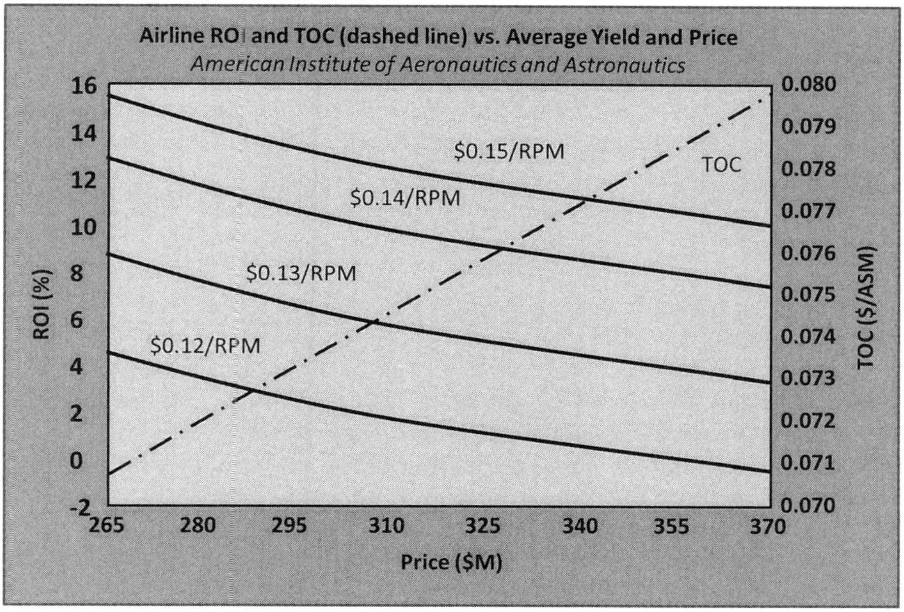

Fig. 2.3 Airline ROI and TOC (dashed line) vs. Average Yield and Price.

2.4 Identifying Manufacturing-Related Costs

The cost to manufacture an airplane is reflected in all the elements leading up to the actual touch labor to produce the airplane. For example, the overhead combines all elements to operate and maintain the factory and business operations of the firm. It also includes the research and development that has been completed or is ongoing as part of the effort to develop the existing aircraft or develop new aircraft for sale.

However, the manufacturer cannot control certain factors external to the enterprise. For example, the number of aircraft ordered, the times of the orders, and the corresponding payment schedules, interest rates, and projected inflation rates are variables the manufacturer cannot control.

The fixed sales price of an airplane has a projected ROI for nonrecurring cost included in its overhead. Every airplane sold recoups a portion of the cost to construct the means of production necessary to build and maintain an aircraft production line.

All costs are captured in the hours to build the airplane. Every hour a mechanic expends to build an airplane is burdened to include the planning and conceptual design, preliminary design and system integration, detail design and development, manufacturing (recurring and nonrecurring) and acquisition, factory operation and support, and storage and disposal costs.

The first step for identifying cost saving is to evaluate the process or activity that contributes to the largest expenditure of touch labor (efficiency).

The second step is to determine if modifications can be made to reduce the number of hours expended without negative impact to the quality of the process or activity.

The third step is to evaluate whether the improvement will affect the burdened hourly rate for each hour expended (impact).

2.4.1 Efficiency

Included in efficiency are all the touch labor hours necessary to make an airplane carry the burden of all other activities in support of its manufacture. This includes purchased materials, parts, and components. The cost to purchase the materials, parts, and components are negotiated and fixed by competitive bid. The continual evaluation and investigation of lower-cost suppliers is continuous by internal departments tasked with monitoring and improving supplier cost and quality.

The greatest opportunity to reduce cost that affects product price is to reduce the internal labor hours required to produce the airplane. Most airplane manufacturers have become assemblers that acquire parts and components externally. Typically, the engine(s) represent the largest acquisition component at 20–25% of the cost of a jet aircraft; the largest internal cost factor is related to the fabrication and assembly of the airframe structure [2.8].

The parts made internally are divided into two areas for considering cost reduction: fabrication and assembly.

2.4.1.1 Fabrication

Labor expended for fabrication is for making the parts, components, and pieces that are delivered to the factory for assembly into an airplane. Highly efficient machines and systems have been developed and implemented within airplane factories to replace the hand operations formerly used to produce parts. For example, hand trim, mill, and drill operations have been replaced with highly efficient and accurate technology enhancements. Technology improvements have also opened up factory space by replacing hundreds of trim and drill templates and bonnets with machine technology. The improved efficiency and the effectivity of the manufacturing process for fabricated parts have for the most part been optimized and currently represent ~35% of the internal cost of the airplane.

2.4.1.2 Assembly

The assembly centers within aircraft factories are the focal points for collecting all the varied pieces of an airplane and bringing them together to be joined into an airframe. Because the airframe represents 65% < 75% of the cost of an airplane's production, and fabrication activities represent 35% of the airframe's cost, the greatest opportunity to evaluate cost reduction activities for touch labor is in the assembly center. Figure 2.4 shows the cost structure breakdown of labor for fabrication and assembly as percentages of the airframe cost.

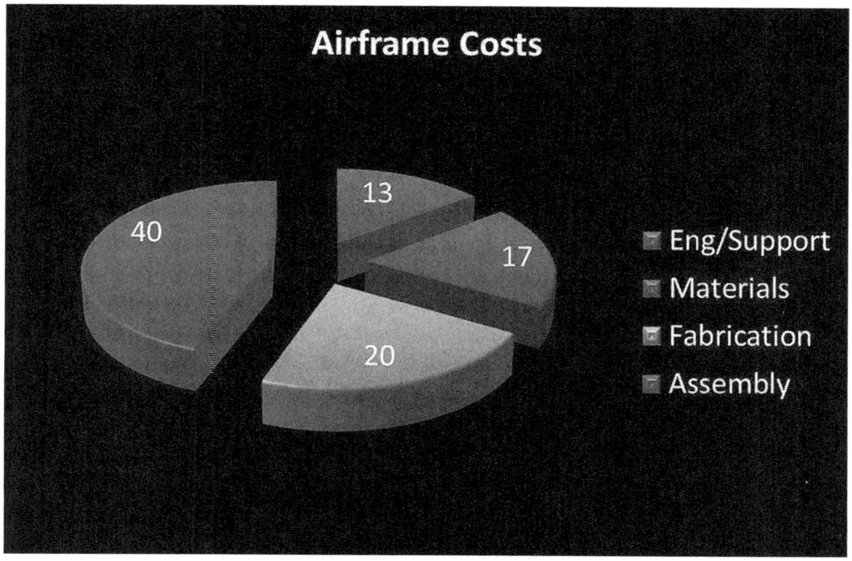

Fig. 2.4 Cost structure breakdown of labor for fabrication and assembly as percentages of the airframe cost.

Assembly centers are where the fabricated parts, components, and pieces are delivered for hand assembly. Hand assemblers place, prepare, and fasten the parts, pieces, and components together. The hand assembly is little changed over the decades since the method of airframe assembly was created. The assembly tool's components are hand assembled into place to position and hand assemble components of the airframe. The components of the assembly tool are then hand disassembled so the assembled airframe can be removed.

The greatest opportunity for cost reduction is resident in the unchanged processes and activities associated with the assembly of the airframe. Therefore, the identification of the cost contributors to determine the opportunity for cost reduction is focused on the assembly center. The major process steps that contribute to the collective cost of the assembly are part placement, fastener preparation (includes D/C), stuffing (installing internal components), and fastening. Further breakdown of direct labor hours to assemble the airframe reveals that 65% of the hours to assemble the airframe reside in the D/C of holes. Evaluation of the process or activity that contributes to the largest expenditure of touch labor (efficiency) is D/C. Combined, the airframe is 65% < 75% of the cost of the airplane; assembly is ~65% of the cost of the airframe; and D/C is ~65% of the touch labor cost of assembly. Therefore, the focus for cost reduction is directed toward activities or technologies that can reduce the labor hours associated with D/C of holes. Figure 2.5 is a pie chart showing the contributors to cost in the assembly of airframes.

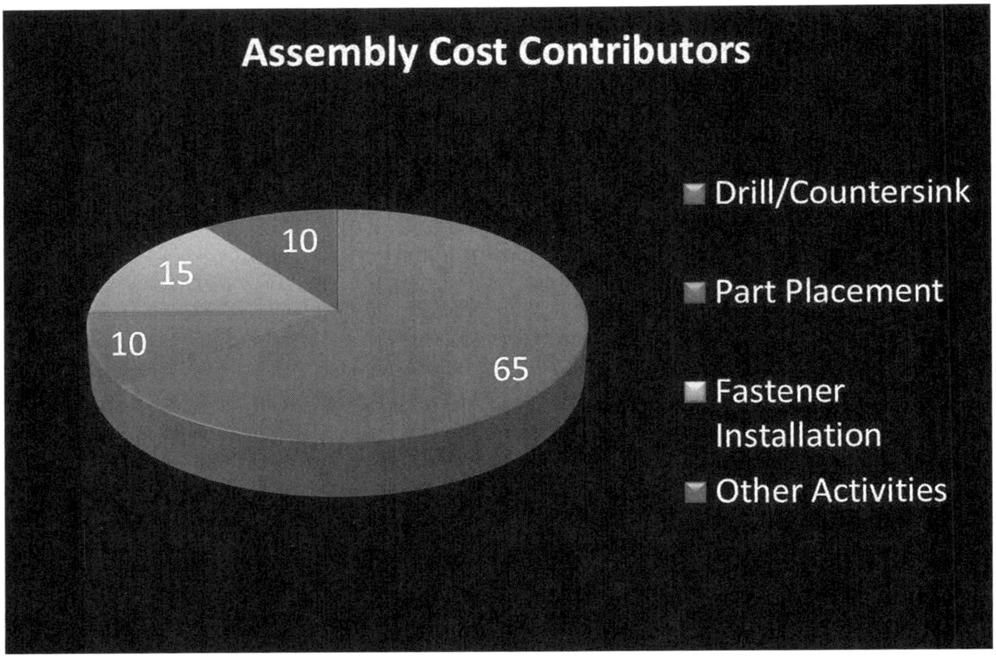

Fig. 2.5 Pie chart showing the contributors to cost in the assembly of airframes.

2.4.2 Effectivity

The determination if modification(s) can be made to reduce the number of hours expended without negative impact to the quality of the process or activity (effectivity) appears simple on its surface. Before any trade study can be done to determine if there are technologies available to improve the hand D/C process, an evaluation has to be done to determine the current effective rate holes are drilled and countersunk. Determining the current effectiveness of the manual D/C process is more complex than just counting holes and dividing it into the hours to drill the holes to get the holes per hour expenditure.

The number of holes drilled and countersunk in an airframe assembly can be large. Over 150,000 holes are drilled or drilled and countersunk in the airframe of the F/A-18 E/F. A 747-400 has six million parts, half of which are fasteners [2.9]. The analysis and assessment of opportunities where new technologies or processes can be applied is performed to determine if and where they can be applied. The study of an airframe for areas where the application of technology might improve the hand method is called a feasibility study. The study is described in Chapter 5, Considerations before Replacing Hand Drilling with Mechanization or Automation, 5.1.1 Feasibility Analysis. The feasibility analysis is performed because not all holes and countersinks are equal. The feasibility study identifies the best opportunity for applying technology and measures the hours to perform the task by hand within the cost center where the holes are drilled and countersunk.

Once the feasibility analysis report is complete, a trade study is performed to canvas the market for technologies that could reduce the labor hours to D/C the specified holes. If the per-hole speed of the technology is greater than the hand method, there is potential for cost savings.

2.4.3 Impact

One of the considerations for the reduction in touch labor hours associated with hand D/C is how to translate the savings to dollars. The reduction of touch labor hours does not remove all of the burdened cost assigned to each hour of labor. A decrease in labor hours may even increase the amount assigned to the remaining dollars because the indirect cost and the costs already consumed in other operations prior to the product build cycle have to be spread over fewer hours. While reducing the number of hours may increase the overall cost of a fully burdened labor hour, there is evidence that some additional benefits derived from the application of automated D/C may have an offsetting effect on the higher burden rate. In some cases, use of automated D/C may even cause enough reduction to offset the fewer hours to allocate indirect, engineering, and development cost. Regardless of whether the offsetting savings reduce the burdened hourly rate, the dollar amount used to calculate savings is derived from a smaller dollar amount. This amount is always a portion of the fully burdened rate to provide a safety margin and to guard against overstating the amount of savings. The per-hour savings calculated in dollars usually runs at the shop rate (mechanic hourly average wage + benefits). Typically, this is about a quarter of the fully burdened hourly rate for touch labor.

Recently, factory leadership has recognized that to understand the cost elements of D/C, they have to break down the tasks, activities, and operations that contribute to the number of hours to D/C of a group of holes within an assembly cost center. Leadership has also begun looking at the cost impact that D/C holes have to other tasks performed in the cost center. For example, the hours allocated and measured to D/C from a group of holes begins when the mechanic logs onto the D/C task. It ends when the finished holes are bought off. In between logging on and logging off, a number of activities are performed that contribute to the cost of D/C of holes, and to other activities within the assembly cost center. D/C of holes creates the need for other activities that cost time and material, and therefore are manufacturing cost contributors.

The entire process from cost center to cost center differs, but a generic description can be created that captures most of the labor activities that contribute to the cost of D/C holes. Most of the activity is non-value-added, meaning the activity is not directly related to creating a hole or a countersink by removing material. If the only activity measured was the material removal portion where the mechanic was either removing the material (called time in the hole), or physically moving to the next hole to be drilled, then the hours needed to D/C a group of holes would be substantially less. Therefore, by studying the D/C support tasks, costs are unmasked that can be evaluated for efficiency improvement. They include moving drill templates or tools to and from the assembly tool; retrieving and installing drills and countersinks in the hand motors; de-burring and/or repairing skins, structure, sealant, and substructure; removing debris from between the skin and substructure; evaluating the pilot hole for position; foreign object elimination; quality inspection of the hole; and quality inspection of the countersink. The quality inspection of the hole is performed separately from the inspection of the countersink. This is due to the fact that hand D/C tasks are performed separately. The hole inspection is also performed three times on military assemblies and some parts of commercial airframes: first when the hole is complete and the skin is attached to the substructure; again in the skin (backside) when the skin is removed from the substructure (called de-stack); and once more on the inside (side against the skin) when the skin is de-stacked.

Cost studies have also been performed to detail the different tasks performed by mechanics while they are in-the-hole. Each event takes time and creates cost. The hand D/C method requires a step process in which first a starting hole (pilot hole) is drilled smaller than the finish hole size. An additional intermediate hole is drilled, followed by drilling of the final size hole. After the hole is drilled and inspected, the countersinking operation is performed shallow, checked for depth, depth adjusted, performed again, and inspected.

All of these activities add cost to the D/C process. When cost savings comparisons are done between automated D/C and the hand method, only time in-the-hole hours are compared. The time in-the-hole savings alone, even at the unburdened hourly rate, is often enough to justify integrating machinery into the process. Automation is used to D/C holes "one-shot" full sized, eliminating the need to step drill. In addition, with automation the D/C operation is performed simultaneously.

Evaluation of the entire D/C process and its effect on manufacturing cost has provided insight into cost contributors to the D/C operation. They include the cost of quality, cost of workforce injury, and support costs (to enable hand D/C).

2.4.3.1 Cost of Quality

Many proprietary studies at major airplane producers have revealed the impact that D/C has on the cost of quality. Some have estimated it as high as 80%. It makes sense. The single largest time (hours) labor contributor is the use of D/C to perform the most "aggressive" tasks (precise removal of material on costly product). The other activities are benign by comparison. Part placement, liquid shim, inspection, test, sealant application, and fastener installation can cause defects. But the single activity that has the greatest effect on the quality of the airplane airframe is drilling and countersinking. Even defects in other tasks within the assembly process are attributable to D/C. For example, a gapped fastener is a fastener defect. The fastener follows the hole. Therefore, the cocked hole caused the gapped fastener, requiring it to be removed and reinstalled.

Another example is a defect in tubes, components, systems, structure, or electrical wiring that is installed internally and covered by the skin. When a drill strikes one of these internal pieces, it causes an impact defect called a drill-start. After the skin is removed, additional inspection is performed to see if drill-starts have occurred. Drill-starts can cause the impacted tube, component, system, structure, or electrical wiring to be replaced. At a minimum, the corrective action involves repair.

Other costly defects that D/C can cause are oversized holes, elongated holes, cocked holes, deep countersinks, shallow countersinks, cocked countersinks, holes below minimum edge distance, and holes in radii. Most of these defects are attributable to the hand D/C operation.

Most defects are caught during in-process inspection. They require reporting, examination, determination of cause, and corrective action. Corrective action results in rework.

Some defects escape the assembly center and are discovered in downstream cost centers or at the customer. A few defects have escaped detection and in rare cases caused failure in the aircraft structure.

The advancement in the field of airframe forensics and technologies to enable the evaluation of catastrophic failures has increased the emphasis on improving hole and countersinking quality. Fastener failure is attributable to the hole and countersink that the hole occupies. Escape defects and even marginally acceptable holes can cause catastrophic failure or reduce an airframe's predictable life.

Many structural failures can be traced to a defective mechanically fastened joint. For example, longitudinal and circumferential fuselage joints are among the most highly loaded parts of an airframe, subject to loading each time the aircraft is pressurized. Over many flight cycles improperly installed fasteners can loosen, permitting microscopic movement between overlapping fuselage panels. Such movement leads to fretting fatigue cracks, resulting in early failure. The highly publicized rapid decompression of Aloha Airlines Flight 243 in 1988 resulted from such fatigue cracks along the fuselage joints. The cracks eventually grew long enough to connect with one another, and catastrophic failure resulted [2.9].

Fastener failure begins with the hole and countersink quality. The spread of automated D/C machines has provided sufficient data to show that they improve hole and countersinking quality while simultaneously reducing defects to Six Sigma level. Specifically, Six Sigma refers to manufacturing processes that produce a level of quality at 3.4 defects per million. The hand D/C tasks result in a rate of 66,807 defects per million, or 3.0 sigma. Commercial airplane manufacturing employees at Boeing drill up to 18 million holes each month [2.10].

Therefore, the inclusion of the cost of quality as a consideration for an automation application ROI as an added cost benefit is being considered.

2.4.3.2 Cost of Workforce Injury

Drilling holes is labor intensive. It involves pushing progressively larger and larger drills through materials by human pressure against a drill motor. The countersink operation is even more stressful as it involves a larger cutting surface that adds the stress of torque to pressure. These operations are done repetitively over the course of an eight- or ten-hour day.

Figure 2.6 shows a mechanic pressing against a drill motor to countersink a hole.

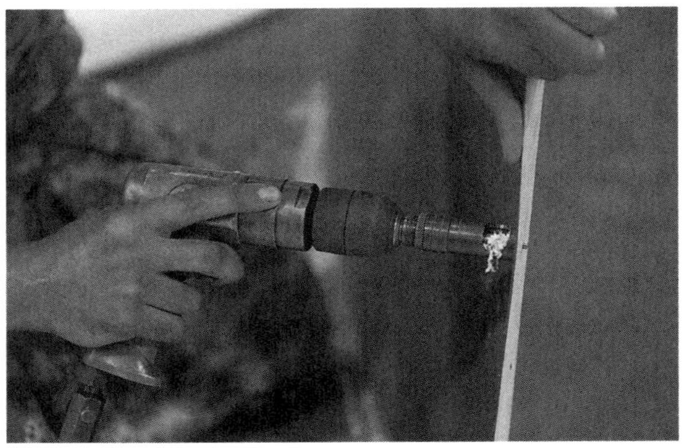

Fig. 2.6 Mechanic pressing against a drill motor to countersink a hole.

The resultant stress radiates to include the back, elbows, wrist, and hand. An aging work force in aerospace has caused the manifestation of injuries resulting from long-term D/C. The average age of a touch labor employee in the U.S. aerospace industry is 45 years old. Twenty-five percent are over the age of 60 [2.11]. Figure 2.7 shows the age distribution of aerospace workers.

The result is that lost-time injuries from one-time events such as a metal chip in the eye or splinter have been compounded by a dramatic increase in repetitive injuries resulting from an aging workforce. The average lost-time injury cost $30,000 (€24,000 as of June 1, 2012). Each injury equals 8–9 lost work days [2.12]. This cost varies by factory, region, state, as well as country. The cost does not include inefficiencies associated with the distribution of the workforce to perform the task of the absent employees. Nor does it capture the cost of the employees' inefficiency derived from light duty (recovery) as they reintegrate into the production workforce.

Within U.S. aerospace manufacturing, repetitive motion accounts for 80% of lost-time injuries, and D/C accounts for 80% of all lost-time injuries.

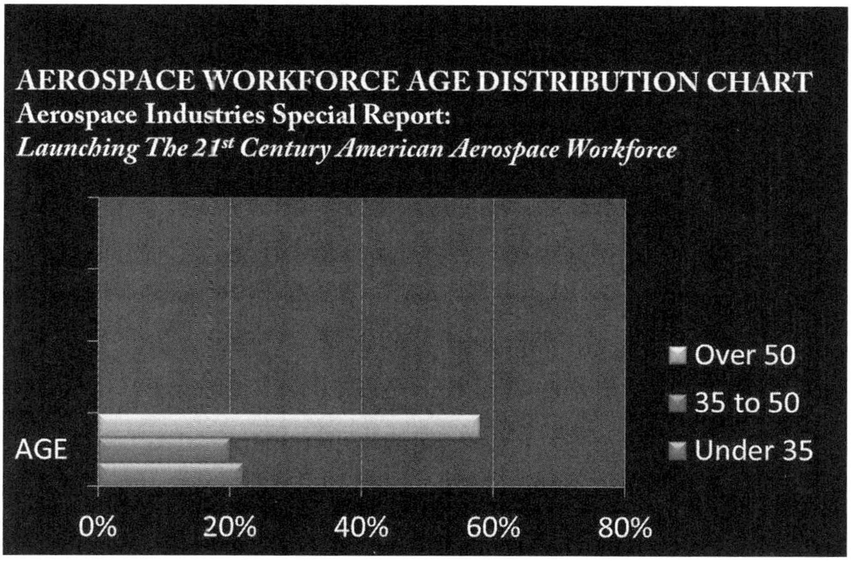

Fig. 2.7 Age distribution of aerospace workers.

2.4.3.3 Support Cost

Support cost directly burdens the touch labor dollar by supplying tools, equipment, supplies, and standard tooling to the mechanic. They include the operation of the compressors that feed the pneumatic drill motors; drill templates and drill tools to position the drills; storage racks to store the templates and bonnets; handling equipment for the drill motors and countersinking cages; tool storage and maintenance cribs for the drill motors and standard tooling; step stools, work stands, and ladders to reach holes; and personnel to support the direct labor for D/C.

Acquiring and maintaining these and other critical components of the hand D/C manufacturing process are other contributors to airframe cost. Figure 2.8 shows a work area cluttered with the necessary paraphernalia to support the hand D/C operation.

These necessary items consume ≈50% of the factory workspace within the assembly cost center. Automation reduces the floor space needed to perform the same operation and lowers the cost per square foot for assembling an airframe.

2.4.4 Cost Contributors Assessment

As more evaluation has been conducted and data collected, the hidden costs of hand D/C have been unmasked. Quality, workforce injuries, and support cost are contributors to the cost of the process.

This data has led to the reassessment of ROI cost benefits contributors beyond simple in-the-hole cost savings.

Fig. 2.8 Work area cluttered with the necessary paraphernalia to support the hand D/C operation.

2.5 Summary

The motivation to reduce price is driven by the ever-pressing desire of customers to reduce price and increase performance. The airplane producers have been caught between the better-for-less dilemmas. Early in the 1990s, the government established and funded programs that promoted design-to-cost initiatives. These initiatives have been largely defeated due to the ever-increasing complexity of airplane production and the demand of the customer, both military and commercial, to improve performance.

Global economics have affected the ability of countries and companies to afford the increasing price of airplanes regardless of performance. The increase in cost reduction efforts for airplanes is driven by the airplane producers to increase revenue by selling more airplanes and maintain their profit margin. To attain this goal, they need to reduce cost to reduce price.

To reduce price, airplane producers have performed analysis to determine the contributors to cost. Once the contributors to cost were identified, further analysis was performed to determine the ability to reduce cost while maintaining or improving the quality of the product.

Analysis has identified the airframe as the most expensive part of the airplane, and airframe assembly as the most expensive part of the internal process that determines the cost of the airplane. Airframe assembly is also the last area of the airplane manufacture that has resisted mechanization and automation improvements. All of the assembly process is performed by hand. Further observation and analysis has shown that the largest contributor to the hours needed to assemble an aircraft is the D/C process. Feasibility studies identify areas where mechanized or automated technology can reduce the cost of assembly.

When applied to the assembly process successfully, automated D/C technology has shown that it not only reduces the touch labor hours needed to produce an airframe, but also impacts the burden added to each assembly hour.

The addition of on-assembly D/C improves quality and therefore eases downstream assembly processes, reducing cost beyond the benefit of producing holes faster. In addition, higher quality derived from automation reduces the cost of quality by defect associated rework. The removal of human beings from the strain of repetitive motion required by hand D/C also reduces the cost of lost-time injuries.

All of the positive benefits of automation result in savings to the airplane producer. The savings are present to increase margin or generate sales through price reduction. While arguably the savings derived from automating the D/C operation are small compared to the cost of the airplane, the small profit margins of commercial airlines make even small savings attractive. The added benefit is lower operating cost as the result of less airframe maintenance due to the higher quality of an automated process.

It should be noted that the cost of holes and the fasteners that fill them has not escaped airplane designers, leadership, and customers. There has been a move in the last decade to replace fasteners and therefore holes with bonded structures. There are problems with implementing the plan: one is opportunity and the other is risk.

The rarity of a new airplane design and therefore inclusion of bonded structures is rare. New airplane designs and production lines are built once a decade. Therefore, the opportunity for inclusion into a new design is rare.

The risk of incorporating a new type of process to replace fasteners is amplified by the fact that crashed airplanes are accompanied by pictures of bodies. New process applications and redesign of the Airbus A300 series where a newly designed vertical stabilizer made of composites and incorporating deviation from traditional fuselage attachment methods has gained infamy. Three Airbus crashes involved a separation of the composite vertical stabilizer (tail fin) from the fuselage. Five hundred, or one in five of the Airbus deaths, including 228 from Air France Flight 447, resulted from these three crashes [2.13]. Airplane designers and customers are very conservative when incorporating design changes to traditional proven standards and processes. The risk of failure might be small, but the incidence of failure causes large repercussions to reputation and cost.

Therefore, in consideration of opportunity and risk, the fastener and the cost to make holes/countersinks necessary to install them will be with us for a long time.

References

2.1 Velocci, A.L., "Profit Wave Uncovers Nagging Paradoxes," *Aviation Week and Space Technology*, May 29, 1995.

2.2 Defense Manufacturing Council (DMC) Executive Group off-site meeting notes, January 19, 1995.

2.3 "Definition of Requirements for an Aeronautics Affordable Systems Optimization Process," proposal to NASA Langley Research Center, submitted by MADIC team, January 1995.

2.4 Housner, J., "Affordable Design and Manufacturing (ADAM) for Airframes and Propulsion Structures," Oral Paper presented to Planning Committee on Competitiveness, NASA Langley Research Center, Hampton, VA, July 1994.

2.5 Greg Reigel, *The Basics of Aircraft Finance*, Federal Aviation Administration, April 3, 2006.

2.6 Boeing Commercial Airplane Prices, http://www.boeing.com/commercial/prices/index.html, Retrieved May 31, 2012.

2.7 Marx, William J., Dimitri N. Marvis, and Daniel P. Schrage, "A Hierarchical Aircraft Life Cycle Cost Analysis Model," Aerospace Systems Design Laboratory, School of Aerospace Engineering, Georgia Institute of Technology, Atlanta, GA 30332-0150, Retrieved May 31, 2012.

2.8 Martin, Rick and Daniel Evans, "Reducing Costs in Aircraft: The Metals Affordability Initiative Consortium," Journal of the Minerals, 52 (3), 2000, pp. 24–28.

2.9 http://www.forensisgroup.com/forensisforum/defective-fasteners-drive-structural-failures, Retrieved May 31, 2012.

2.10 Arkell, Debbie, "Built-In Quality, Better, faster, cheaper—but most of all, better," Boeing Frontiers, Volume 02, Issue 03, July 2003.

2.11 Aviation Week Workforce Study, Contributing partners: Aerospace Industries Association, American Institute of Aeronautics and Astronautics, National Defense Industries Association, 2009, www.ndia.org/Divisions/Divisions/STEM/Documents/ArticlesOfInterest/Aviation_Week_Workforce_Study_09.pdf, Retrieved May 31, 2012.

2.12 Jordan Sikes, Susan, OSHA, Regional VPP Manager, "Achieving Safety and Health Excellence Through the Voluntary Protection Programs, (VPP), 2012, http://www.osha.gov/dcsp/alliances/nsc_airline/sikes_presentation.ppt, Retrieved May 31, 2012.

2.13 Cox, William John, "Ground the Airbus?" Global Research, Centre for Research on Globalization, http://www.globalresearch.ca/index.php?context=va&aid=14025, Retrieved May 31, 2012.

Chapter 3
Incentives and Disincentives to Automate

If you have a rigidly controlled economy, cut off from the rest of the world by infinite protection, nobody has any incentive to increase productivity and to bring new ideas.

—Manmohan Singh

3.1 Introduction

Airplanes are highly complex pieces of technology that have evolved to the extent that only a handful of companies and a handful of countries can produce them in their entirety. In fact, they have become so complex and expensive to produce that even the big name companies such as Boeing and European Aeronautic Defense and Space Company (EADS) are assembly centers for component pieces fabricated by geographically dispersed companies that produce individual parts of the airplane. Much of the geographic distribution is political or business driven. Many times the distributed work content does not optimize the desired manufacturing plan.

New aircraft assembly lines are built infrequently for commercial, business, and private airplanes. They are rare for military programs. Unlike automotive assembly lines where several new lines are built each year that incorporate manufacturing enhancements, airplanes are assembled at very low rates and have long lives. Change or enhancements to their operation over their 20 to 30 years lifespan comes with great risk to the airframe assembler. The airframe assembly line becomes "settled" with robust processes that can be repeated to produce acceptable quality airframes. Change activity has reached its minimum level of disruption. The product has reached a price point that is acceptable. While the customer desires a lower price, they do not want it at a loss of performance characteristic such as fuel economy (commercial) or stealth (military).

The airplane is also produced by companies and countries that have evolved a production process to a state of understanding. They are "the few" who know how to build an airframe, and any change to their infrastructure, labor relations, political influence, and processes developed over decades threaten the livelihood of the company.

The parts are expensive to fabricate and limit the opportunity for test pieces to incorporate automation rapidly into an existing production line with lower risk. Risk includes labor and political impact. It is the rule rather than the exception that the first assembly that applies automation technology must address all issues including geopolitical pressures, labor, and business operations as well as technical challenges.

Therefore, the disincentive to apply automated D/C or any type of automation is challenged by the age-old adage, "If it ain't broke, don't fix it!"

This is the initial challenge the automation engineer must overcome. Once overcome, then a scientific and calculated approach to automating the largest cost driver (D/C) can be performed to determine the legitimate challenges and incentives to automate. Viable and objective data can be obtained to determine the incentives and disincentives to apply automation into a production line.

This chapter provides the elements of consideration when approaching the decision to apply automated D/C onto a production line. There are mature assembly lines and opportunities within fully functioning lines with a minimum or no disruption to the existing line flow. There are also opportunities to integrate automation into the plan for new airframe assembly lines.

This chapter will also provide considerations and information to assess the extent to which a new airframe assembly line should introduce automation and how to determine what impact automation will have on non-technical issues such as geopolitical considerations and labor. The decision to automate comes with the responsibility of judicious and cautionary application decisions with carefully targeted areas of the airplane for installation.

It also involves evaluating all the pressures that influence its application. A successful first installation will incentivize expansion.

The incentives and disincentives for applying automated D/C are different for the military, commercial, business planes, and space launch vehicles.

3.2 Military Automation Disincentive

The military approach to acquisition of airplanes is quite different from that of commercial airplane buyers. The military pays for the means of production such as tools and test equipment. Commercial airplane producers pay for their own means of production. Military airplane customers (includes helicopters) buy small lots that are subject to political winds. Most often they will fund annual buys and occasionally multiyear buys with short time horizons. Commercial airplane producers build production lines that reflect the market prediction into longer time horizons. Military acquisition teams demand tight controls over pricing and profit. Commercial airplane producers are incentivized to maximize profit and establish price points that reflect their desire to sell and make money. Military suppliers are restricted to exporting their technology to government-approved buyers, if at all, whereas commercial airplane producers can and do sell to whoever can find the money to buy. Military airplanes in the United States and Europe do not have competition. Once the contract is let, they are the source for all buyers. There is competition in the commercial market, and even though options are restricted, a commercial airplane can be bought by airlines from a number of sources.

In military acquisitions, production tooling, method of manufacture, lot quantities, restricted sales, restricted profit, and single-source procurement disincentivizes the application of automated systems for production.

3.2.1 The Means of Production

The growth of the airframe assembly line up to the beginning of 2000 was built (and is mostly built today) to facilitate human beings interacting with large steel boxes. The boxes are constructed to hold the assembled pieces of the airplane they are custom designed to build. The finished subassemblies are mated together into progressively larger assemblies until an airplane emerges. They are assembly tools that grow in size as the airplane they were meant to build grows until the assembly tools weigh many tons. Additional tooling includes master models, master gauges, templates, drill bonnets, handling fixtures, holding fixtures, and dozens of other tools meant to aid and facilitate the assembly of an airframe. Most of the large structural tools are oriented to build the progressively larger sections of the airplane in flight orientation. They are built with their bellies down and elevated off the ground so mechanics can get underneath them and work. As a result, the large complex fixtures are surrounded by work stands to enable mechanics to access the elevated airframe subassemblies and assemblies. Figure 3.1 illustrates the size and complexity of a military assembly tool.

Test fixtures and special test equipment are built for system, continuity, and leak checks as the subassemblies grow and mature. The progressively larger subassemblies are lifted out of their boxes to the mate tools by crane. A modern military airframe can require hundreds of crane moves to facilitate the assembly of the airframe.

Fig. 3.1 Size and complexity of a military assembly tool.

Tools flow into and out of storage centers. It is estimated that a major assembly center such as Boeing, Lockheed, or Northrop Grumman manages hundreds of thousands of assembly related tools for various programs. Tools are duplicated to maintain production line continuity when tools are misplaced, damaged, out for calibration, or maintenance. Tool control departments manage the flow of the tools in a coordinated dance of demand and supply to meet the needs of the product delivery teams.

The hardware in this method of manufacture is so rigid and product specific that it is highly change sensitive in a product that is so complex that its infancy is fraught with change. A fluid design with thousands of changes in the first one to 100 units, wreaks havoc with an assembly line built to be inflexible to change. Large assembly tools have been and will continue to be chopped into pieces and lengthened by a mere 6-in change in length of the airframe.

All this hardware is supported, maintained, managed, and moved by a complex multilayered deep-dish bureaucracy of human-ware and software. The bureaucracy is in place to collect and offer up proof of engineering, cost, and schedule conformance. To be sure, the method produces good airframes. But is there another way? All of this is paid for by the defense department, including an allotment to replace aging and lost tools.

The determination and justification of how much an airplane costs the U.S. Department of Defense is justified and substantiated by the method of its manufacture. The military wants performance and looks to the manufacturing experts such as Bell, Boeing, Sikorsky, Lockheed, BAE, and others to build the airplane.

When a machine is purchased that could improve airframe assembly, it is capitalized. It is paid for directly out of the company margin, and it affects the company's market position by lowering its return-on-asset. The tools that the machine replaces are lost revenue because all of the tooling is burdened, including profit. The savings from the improved assembly method (reduced touch labor) reduce the price of the plane because the savings are passed onto the buyer.

So where is the incentive to change? The government pays for the cost of production. The government pays for a highly complex hand-built method of manufacturing. Any reduction in cost (savings) is passed on to the customer. And the government wants the manufacturer to pay for the machine that will reduce its revenue and profit.

A few past programs have attempted to break this cycle by offering to split the first year or first several years' savings with the company for manufacturing improvements. But the incentive to split a few dollars today does not make sense when the manufacturer that does not implement savings measures can keep all the money over the life of the program.

There are some individuals within government procurement that are beginning to recognize this dilemma. They are exploring ways to differentiate machinery operated specifically for one program from machines that perform operations on several programs. This means that the product-specific machinery can be purchased as assembly tooling and therefore paid for by the government. It thus removes part of the disincentive to improve production assembly by applying automation.

3.2.2 Parceled Application

When automobile manufacturers build assembly lines, they commit dollars to construct an entire line based on the maximum rate (units per hour). The auto assembly line is built to maximize efficiency at the maximum rate of production.

When a military production line is built, the tooling is designed and constructed to build the first few units over several years. They are designed for initial low-rate production, say one or two a year, even if the projected number of units over the life of the program is in the thousands. As each progressive block of airplanes is released, the number of orders increases. A rate tool analysis is performed at each block order release to see how many more of the tools are needed to meet the new rate. These are called rate tools. This philosophy of parceling the application of tools for limited use and replicating the same tools over and over as production quantities increase results in the least efficient means of production. Tools meant to produce a small number of units are replicated at government expense to aggregate to larger quantities of product.

The replicate rate tool philosophy prohibits the acquisition of highly efficient tooling, mechanized systems, or automation that would maximize efficiency at higher rates. The added expense of the systems in the low-rate initial production phase would increase efficiencies over the life of the program. The myopic tooling approach is driven by annual incremental funding and block release of orders.

3.2.3 Specific Use

The government prohibits the cross program use of their property such as tools, machines, and test equipment. A piece of test equipment purchased with F-35 funds to test for leaks in a fuel tank cannot be used on say an F18, even if the test equipment is used infrequently.

This specific use procurement law extends to every aspect of production. A piece of software purchased by one program and installed on a company-owned machine, to maximize the machine's accuracy, must be turned off if any other program parts are milled, trimmed, or drilled on the machine.

Therefore, if the government were to modify its approach to buying the means-of-production to machines and automation, the machines or automation would sit idle during the low-rate initial production even if another program's activities could be performed.

Another program (or the company) cannot take advantage of the available machine or automation, even for a minute. In fact, a clear segregation rule is required that specifies that aisles or fences be built to maintain separation of one program's property from another.

The specific use and separation requirements depress the incentive to buy highly effective machines, mechanization, or automation.

3.2.4 Time Horizon

When a production line for an automobile is built, the goal is to get in and get the line running as fast as possible. Time is money. Money does not flow until the line ramps to maximum efficiency.

Military production lines are built in blocks over time horizons that extend out ten years or so. This blocked approach to build production lines incentivizes a short-sighted approach to goals. For example, the release of an initial block of ten airplanes due in 2.5 years with a completion bonus when successfully completed on time drives a ten-airplane goal. The focus is to build ten airplanes on time, not the potential to build thousands. Therefore, the military assembly line grows in little increments with monetary and leadership rewards for completion. Acquiring and installing machines, mechanization, and automation requires longer time horizons and vision. A director becomes a vice president because he meets the goal. The next director meets the next goal. It is a short-sighted recipe for program efficiency, product quality, and effective application of resources.

3.2.5 Politics

All products today are manufactured with political consideration. Military and commercial airplanes are no exception. Military airplanes are especially subject to the whims of political pressure because of the government's direct participation in the procurement process. This affects the method and efficiency for an airframe's production.

Airframes have logical production breaks where the individual components are sub-assembled as small units are progressively combined until they become a finished airplane. It is most efficient to maintain the subassemblies close to their collective final assembly points to control precise mating that occurs when they are joined together. Assembly centers for airplanes represent very large investments, and the nature of the assembly process creates a factory that is immobile for the final assembly process.

Political pressure derives from senators and congresspersons who sit on powerful committees that determine the annual fate of defense programs. Companies are well aware that if they place work into states that these senators or congresspersons represent, the likelihood of funding a new program or the continued funding of an established program is certain.

This political consideration subjugates efficient and effective manufacturing consideration to the piecing out of the airframe to diverse geographical locations that may or may not have an aerospace trained workforce. The political parsing of the airframe adds cost by increasing the infrastructure travel, manage, and control of parts and pieces so they fit properly at assembly.

The breaking of the airplane into pieces also defeats the application of machines, mechanization, and automation for other reasons. Two of the reasons are ROI and jobs.

3.2.5.1 Return on Investment

The ROI required to justify a piece of equipment such as automation is derived from its ability to increase efficiency over the current production method sufficient to meet or exceed the cost of its acquisition. The more the airframe is broken down into pieces, the smaller the opportunity to pay back an investment. The government pays for the less-efficient hand method. Rate tools are made at the home assembly center, and manufacturing teams are sent to the state and company where the parts will be made. The price and cost of the product grows.

3.2.5.2 Jobs

The primary reason that powerful senators and congresspersons want work in their states is jobs. Obviously, jobs equal votes, which result in the retention of the senator or congressperson's job.

Therefore, any suggestion of cost-reduction efforts that would decrease the number of jobs resulting from machines, mechanization, or automation is frowned upon.

3.2.6 Controlled Profit

In a free market economy, companies are incentivized to reduce cost (increase efficiency) to increase profit. They are also incentivized to improve quality to increase sales (revenue), further increasing profit.

In military acquisitions, companies are profit restricted. That is, they are limited to a small profit margin due to the "tax payer" effect. The tax payer is funding the project, so profit should be restricted to a small percentage of revenue. This profit-limiting rule has a contravening effect on the cost of product to the buyer.

For example, if I sell a dollar worth of product and I am restricted to 5% profit, then I receive five cents profit. But if I can justify that the cost of manufacturing my product will result in a sales price of two dollars, then I get a dime profit.

Military procurement works this way. The more difficult and complex the means and method of production, the more costly the product is to produce. The more costly the product is to produce, the more the customer can be charged. As long as the manufacturer can justify the manufacturing method and cost, the result is higher profit. Hand building airplanes maximizes profit.

Any reduction in cost (increased efficiency) results in two automation application disincentives directly attributable to military acquisition, profit and jobs.

The cost/price dance that takes place is derived from the price sensitivity point of the customer. Push too hard, and the program reaches the price cancelation tolerance threshold. The B-2 and the F-22 are examples of the manufacturer exceeding price tolerance threshold, where these programs were canceled before reaching their projected production life. Figure 3.2 show the B-2 on the assembly line in Palmdale, California.

Production cost of these high-performance aircraft was eventually reduced from 132 to 21, of which 20 are in active service. The production cost of these aircraft rose to the prohibitive unit cost of $1.07 billion each in 2012 dollars [3.1].

3.2.7 Restricted Sales

Because of the highly sensitive components of military airplanes and nature of their purpose, governments restrict their sale. There are examples of export sales of military airplanes such as the highly successful F-5 or the C-130. Even performance airplanes such as the F-16, F-35, and Bell AH-1 helicopter have been exported. The complex nature of the export sales process, the price of the product, and the function of the product limit the extent of their sales and marketing and therefore reduce the numbers that could be produced.

Fig. 3.2 B-2 on the assembly line in Palmdale, California.

Low numbers of units sold internationally are usually comprised of airplanes nearing the domestic end of their production life. Therefore, the international sales units are integrated into the rate tooling to fill gaps left by the declining domestic orders. As a result, the opportunity for efficiency enhancements (automation) does not exist as a result of a surge in sales (typical in commercial), but as add-on piecemeal sales to extend production.

3.2.8 No Competition

Limited product choice exists in the commercial world. Little or no product choice exists in the military airplane world.

A look at the commercial (including business and private aviation) world shows that there are choices for the operator of the airplane. Falcon Jet or Gulfstream; Boeing and Airbus offer choice, all during an airplane user's operating life. An operator can choose to buy more of a Boeing airplane or change to an Airbus variant if the price and opportunity are right.

Once the choice is made for a specific military piece of hardware, the commitment is made for the life of the program. The numbers may vary or the program could be canceled, but the choice is made.

3.3 Military Producers' Incentives to Automate

After reading the disincentives to automate a military assembly line, it would take a delusional optimist to attempt any installation of a machine, mechanized system, or automation. Apparently there have been some delusional optimists running around the halls of government and aerospace companies that have planted the seeds of change. Some of these seeds have taken root. Evidence of their influence can be seen on the Northrop Grumman's Palmdale production line for the F-35 Lightening II. Figure 3.3 shows part of the F/A-35 Palmdale assembly line.

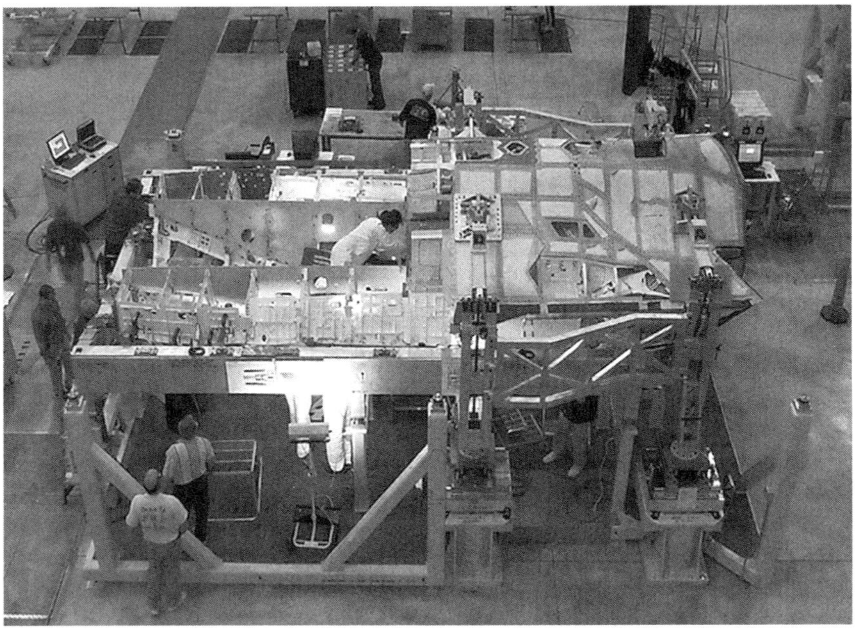

Fig. 3.3 F/A-35 Palmdale assembly line.

3.3.1 Affordability

Cost and the price of a military airplane have opened some eyes to the production methods currently being used. Affordability without compromise to performance has caused a reevaluation of the entire process for funding and manufacturing airplanes. This has led to some changes that have enabled (incentivized) a longer time horizon approach to funding the means of production and the government funding of automation as a tool for assembly.

3.3.2 Enlightenment

A new group of government procurement officers is viewing mechanization and automation differently than the previous generation. They do not look upon its application as risky, but as a facilitator to the production process. As a result, they do not ask why we want to apply automation to the production process, some are asking why we are not applying automation.

3.3.3 Complexity

The complexity of modern airframes and the precision needed to fabricate and assemble them has outpaced the ability of humans to meet the demand. The military airframes of today have increasingly incorporated an imprecise material (composite thickness control) into an airframe that demands "as-manufactured" stealth characteristics. The stealth characteristics of minimal steps and gaps between panels and skins of the Outer Mold Line (OML) of the airframe have to be delivered on-assembly. The B2 experience of filling and sanding the final airplane into stealth-conforming step/gap specification was prohibitively expensive.

This on-assembly process requires machine precision. This is only one of the areas of modern military aircraft assembly that is driving processes away from the hand methods.

Another area that is driving automation is the application of Low Observable (LO) coatings on military airframes. The precise spray application of a toxic radar absorbing material (RAM) [3.2] is best suited for robots. While its application does not apply to auto D/C, its use familiarizes manufacturing, facilities engineers, and factory personnel with the operation and presence of automation in the factory.

3.3.4 Multiyear Buys

Annual buys have always been an inhibitor to the acquisition and integration of automation onto the assembly line. They cause negative ROI assessment for the acquisition of assembly machinery, mechanization, and automation.

I have been told that I had to show an ROI within one year (an impossibility), because that was all the airframes that were funded. This was done knowing that each year would with all probability result in additional airframe buys.

This is another area that has been at the forefront of government procurement attention for decades. Within the last decade, the economic benefit of multiyear buying has emerged and entrenched bureaucratic procurement policies overcome in the interest of affordability. Subcommittee on Tactical Air and Land Forces extended the Navy's F/A-18E, F, and G aircraft multiyear procurement authority for a fifth year [3.3].

3.3.5 Quality

Early and growing application of on-assembly automated D/C machines demonstrated their ability to dramatically improve the quality of the elements (holes and countersinks) and the product. As a result, the quality incentive grew as an independent motive to applying automation. The quality motive was embraced as a feature that could be used to justify and integrate automation without any corresponding decrease in cost that would affect profit. It demonstrated the intent of the manufacturer to improve processes to the customer.

3.3.6 Marketing

This feature is a valuable tool to promote the advancement of manufacturing technology for the betterment of the product. The performance-driven customer appreciates and embraces technology that enhances the reliability and performance of the product, especially if it is promoted as advanced tools to build advanced military hardware. Therefore, marketing the manufacturing as well as the product technology has a symbiotic relationship for program promotion.

The relationship is an incentive to expand automated D/C technology.

3.3.7 Access

Another incentive for automating D/C military and commercial airframe assemblers is access to airframe structure. For example, the F-35 bifurcated inlet duct has hundreds of holes that need to be done using D/C in a very confined space. Figure 3.4 shows the difficulty a human would have to access and perform the D/C operation.

When humans perform D/C in confined spaces, defects increase, performance decreases, and ergonomic-related injuries increase.

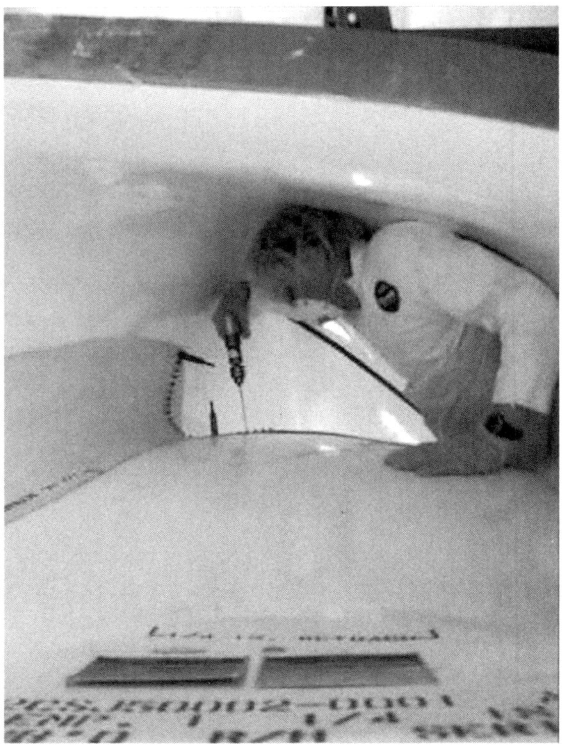

Fig. 3.4 Difficulty a human would have to access and perform the drill/countersink operation on the F-35.

Figure 3.5 shows the automated solution to perform the same task with automation. Ergonomics and the difficulty of the operation have incentivized automation as an alternative.

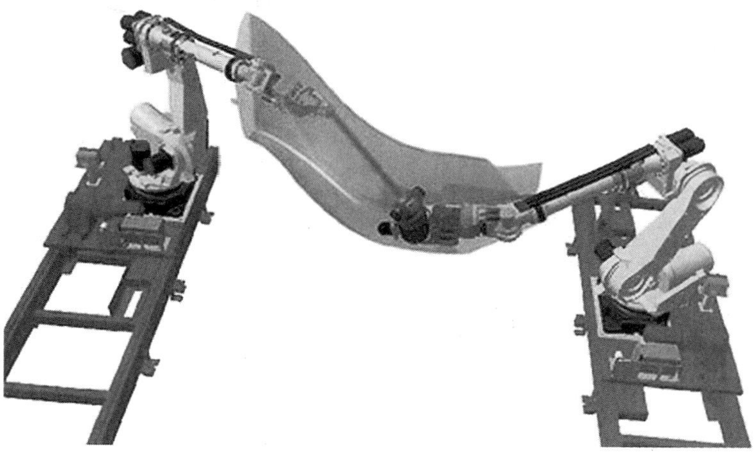

Fig. 3.5 Automated solution to perform the same task.

3.4 Commercial Disincentives

Commercial airframe manufacturing labors under fewer disincentives to automate their assembly processes than military airframe producers, but they do exist. Two of the primary disincentives are geopolitical and labor.

3.4.1 Geopolitical Local

There are pressures on Boeing and Airbus from inside the United States and the European Union to distribute work domestically. Localized production of an entire airplane is no longer a viable political option to sustain production. The complexity and variety of commercial airplanes have caused the distribution of highly specialized work to various regions of the United States for Boeing, and the European Union for Airbus. The political considerations for work distribution mirror those for military aircraft. Airbus and Boeing have to consider the political implications of their decisions for the entire product line, when the mix they are producing is comprised of military and commercial. Therefore, the same influence can be obtained from a senator, congressperson, or government representative by placing work into their state, province, or region. The work placement may favorably influence laws and regulations pertaining to commercial aviation or it may influence support for military programs. The senator, congressperson, or representative does not typically care what type of jobs and work are placed into their state, they just want the jobs.

The geopolitical distribution of work inhibits automation application in commercial programs the same way that it inhibits it in military programs. The greatest disincentive for automation is the reduction of manual labor (jobs).

Economic development teams incentivize companies from outside their state to come and set up operations in their state. The amount of incentive depends on the type and quantity of jobs the company will bring to the state. High-paying prestigious jobs such as aerospace substantially increase incentives such as tax breaks and training. People pay taxes and vote. Robots and machines do not.

3.4.2 Geopolitical Global

The global marketplace has resulted in the manufacture of airframe components and assemblies internationally. Governments such as China and India do not just want airplanes, they want jobs and technology. Therefore, to sell commercial airplanes, Boeing, Airbus, and others have to give pieces of the airplane to other countries to build. They also have to teach the other countries how to build them. As a result, components and subassemblies are delivered to the Boeing and Airbus large assembly centers in places such as Washington and Toulouse, France for final assembly. The low-cost labor and the need to provide jobs in the countries supplying the parts defeat the incentive to automate large portions of the airframe. A globally distributed supply chain resident in low-cost labor markets with high unemployment defeats the desire to automate.

Governments, particularly in developing countries where unemployment and under-employment are a problem, find the justification of automation difficult to accept.

3.4.3 Labor

When new automation technology is available for adoption, the question of its effect on employment is raised. The consequences of automation have been a topic of discussion and debate for decades. The introduction of automation to replace manual effort is a sensitive issue for management and labor [3.4]. Labor's view of automation is driven by its perceived impact on current and future jobs. Automation is not a new issue between management and labor. Articles have been regularly written about the subject since the mid 1800s. Labor's view of automation has been tempered over the decades by reduced human effort, shorter work weeks, and improved wages. A large contributor to the positive effects on labor was the result of negotiated concessions by management in order to install machinery, mechanization, and automation [3.5]. Cost savings derived from automation has enabled concessions such as higher wages, shorter work hours, and less strenuous work while maintaining healthy profit margins.

Automation is a confusing issue for labor. On the one hand it improves the human condition in the workplace, and on the other hand it reduces jobs.

The local and international geopolitical influence of labor can have an adverse effect on any attempt to reduce cost. The National Labor Relations Board's lawsuit against the aerospace giant, Boeing, says it retaliated against its unionized workforce in Washington State by opening a new production line for its 787 airplane in South Carolina, a right-to-work state. The agency (NLRB) wanted a judge to order Boeing to return all 787 assembly work to Washington, even though the company has already built a new $750 million South Carolina plant and hired 1000 new workers there [3.6]. Boeing was incentivized to open a 787 assembly center in South Carolina to reduce cost derived from employing lower-wage workers (hour and benefits), and increase the efficiency of the assembly line by application of advanced manufacturing technology such as automation. As a result of the dispute, Boeing will operate two assembly centers, one in Washington and one in South Carolina. Figure 3.6 shows the Everett, Washington assembly center for the 787 Dreamliner.

Fig. 3.6 Everett, Washington assembly center for the 787 Dreamliner.

3.5 Commercial Incentives

There are many opportunities to introduce automation into a commercial, business, or private airplane assembly line. Commercial, business, and private airplane assembly lines are built more frequently than military assembly lines. The frequency enables the infusion of fresh technology into the production processes rather than squeezing technology into a process designed to operate without it.

The main incentives to apply automation are cost reduction and quality improvement. In the commercial, business, and private airplane manufacturing business, cost savings can be retained as profit, or passed on to the customer as price reduction. Price reduction stimulates sales (raising revenue) and customer satisfaction.

Organized labor (most aerospace assembly line labor is organized) can and does make concessions to allow automation. Automation offers advantages to labor that should be articulated and negotiated to desensitize labor's concerns. In the past, labor has permitted increased automation, with a percentage of the savings from automation going to wage increases. Cooperation and shared advantage can assuage the concerns of labor and further incentivize the spread of automated systems.

3.6 Summary

Incentives and disincentives for applying any automation on an airframe assembly line exist. Military and commercial, business, and private airframe producers encounter different disincentives and incentives resulting from customer requirements and business operating environments.

Military requirements and the business operating environment emphasize performance and affordability while governing the acquisition of product in a rigid and inflexible system that reduces incentive to improve manufacturing processes.

Many factors restrict the freedom to make change in the military manufacturing process. They include annualized procurement subject to political change, geopolitical considerations, bureaucratic and cumbersome regulations, payment methods, restricted profit, and limited time horizon goals.

The disincentives seem overwhelming, but champions of change have created windows of opportunity. Change to annualized procurement, payment methods, and manufacturing awareness have contributed to new technology making inroads into the military assembly process. An example is the new F-35 assembly line in Palmdale, where KUKA Robotics is a key automation contractor on the F-35 Lightning II Program. Previously, KUKA had been a major automation supplier to the automotive industry. The use of automotive automation experts to infuse automation into military assembly lines reflects the changing approach to military acquisition. Figure 3.7 shows a KUKA robotic system on the F-35 assembly line.

Commercial, business, and private airplane producers have fewer restrictions (disincentives) to incorporate cost savings measures such as automation into the production process. They also introduce more products and, therefore, have the opportunity to apply automation into new assembly lines more frequently than their military airframe counterparts. They are also able to apply cost savings to profit, or offer reduced price sales incentives to customers without affecting profit.

Fig. 3.7 KUKA robotic system on the F-35 assembly line.

Within both military and commercial, business, and private airplane manufacturers, labor relations plays a large part in changes and planning to operations that are perceived to affect the jobs-related content of work. Many of these concerns can be elevated to highly charged political events played out on the public stage. Occasionally, global and local geopolitical forces combine with labor to form a collective event providing drama to the press.

All of the disincentives have been played out for decades and eventually been overcome by the cost benefit as well as the benefit to labor that automation provides.

As Aristotle said millennia ago, "When looms weave by themselves, slavery will end."

References

3.1 Consumer Price Index (estimate) 1800–2012, Federal Reserve Bank of Minneapolis, http://www.minneapolisfed.org/community_education/teacher/calc/hist1800.cfm?, Retrieved May 30, 2012.

3.2 Toxic Torts, http://www.elllaw.com/cm/Walter-Lack-Publications/Personal-Injury.html, Retrieved May 31, 2012.

3.3 FY 2013 HASC Subcommittee Mark-up Details, Subcommittee, Tactical Air and Land Forces, http://www.afa.org/PresidentsCorner/WashingtonPerspective/2012/FY_2013_HASC_Subcommittee.pdf, Retrieved May 31, 2012.

3.4 Stern, James L., "Automation-End or a New Day in Unionism," The Annals of the American Academy of Political and Social Science, November 1963, Vol. 350, No. 1, pp. 25–35.

3.5 Beecher, Ed, "Automation and the Trade Union," From *International Socialist Review*, Vol. 23, No. 3, Summer 1962, pp. 87–90.

3.6 "Boeing labor dispute turns into headache for Obama," Posted June 29, 2011, http://www.live5news.com/story/14997872/boeing-labor-dispute-turns-into-headache-for-obama, Retrieved June 1, 2012.

Chapter 4
Types of Drill/Countersink Automation

The strongest principle of growth lies in the human choice.

—George Eliot

4.1 Introduction

There are many types of automated D/C machines that have emerged to fill the demand from airframe manufacturers. They range from small, pneumatically operated, two-axis machines with simple PLCs to the most sophisticated articulated-arm robots positioned by laser with highly precise multifunction end effectors.

Knowing the range of machinery and the types of controllers available provides the options necessary to make the correct choice. Not only will this chapter provide descriptions, capabilities, images, and examples of applied and emerging automated drilling machines, it will evaluate and rank the performance of the machines and their producer. It will also provide information and feedback from the users of the machines, where they are applied.

The variety of machinery available today and emerging for tomorrow provides choices that complicate the selection process. Not only is the performance of the machine a consideration but also the ability and reputation of the company that stands behind the product. Performance alone cannot compensate for a rapidly degrading piece of machinery that has to be constantly maintained, or a piece of machinery that is installed but is prohibitively expensive to program and operate.

Today's automation specialists in the machine industry approach the application of automated drilling machines differently from the purchase of a machine tool by a user. The unique nature and specific purpose of the automation produced to D/C holes require that a partnership for success be built between the builder of the automation and the user. The symbiotic relationship is derived from the unique design for each piece of machinery. Therefore, the machine, unlike off-the-shelf mills, lathes, and drills, is custom built from mostly COTS components that are combined in a unique way to meet the highly specialized requirements for its application. This means that not only should the choice for the piece of automation designed for your specific use be functional, but the company producing it must be sustainable and therefore capable of maintaining the equipment they produce.

The choices are many, and the success of the machinery you choose is grounded in both the proper choice for your specific need as well as the marriage of your company with the automation supplier who should be your partner through the installation, integration, and operational phases of the process.

Buying a piece of highly customized automation while cutting corners by assuming you can integrate the technology into the operation of your factory without the aid of the supplier is a recipe for failure. It has rarely been done.

Therefore, this chapter will also include images and descriptions where the user and the supplier have strategically allied themselves for success. This chapter will also provide description and images of some very good boat anchors where the technology and machinery were sold and bought as stand-alone technology without consideration of an integrated supplier-user relationship. The right choice can sometimes be derived from a bad example. In some cases these failed examples were recovered, dusted off, and effectively used in collaboration with a supplier-user relationship for another application. The basic elements that describe the requirements for an automated D/C machine are motion, action, control, integration, and sustainability.

4.2 Motion

Motion is defined as how far a machine must travel to meet the requirements needed to perform its tasks. Size, location, positional accuracy, and repeatability are the starting point for the physical definition of a machine's motion envelope. The distance a machine must travel and its reach (its envelope) influence its design to meet the defined engineering requirements (hole position). The placement of the machine and interaction with its environment (location) also influence machine design considerations. The primary consideration for machine motion that size influences is the positional accuracy and repeatability. The dynamics that affect the positional accuracy and repeatability include size, reach, load at the end of reach, stability, vibration, temperature variation, and resistance. For example, a small machine in close proximity to its work-piece that drills holes in soft material (little resistance at the tool tip) is designed differently than a large machine that must reach out to drill around a curved surface through tougher material such as titanium. Finite element analysis provides valuable data to design automated machinery to meet the requirements of the process. With finite element analysis, the behavior (stresses, strains, and displacements resulting from load conditions) of large-scale structures is approximated by a Finite Element (FE) model consisting of structural elements (members) connected at structural node points. Calculations on traditional computers are performed at each node point and results communicated to adjacent node points until the behavior of the entire structure is computed [4.1].

4.2.1 Size

Size determination begins with two basic steps.

Step one is to determine the X and Y travel of the machine in order to reach all the part's holes that have been identified for D/C. Included in step one are additions to the basic X and Y determination. Overage determination to the X and Y envelope must be made to include requirements for a parking space, tool changes, clearance, and part removal. Included in the X and Y size calculation is the pitch the machine will need to access curved structures. Figure 4.1 shows a diagram of the basic X and Y machine motion determination.

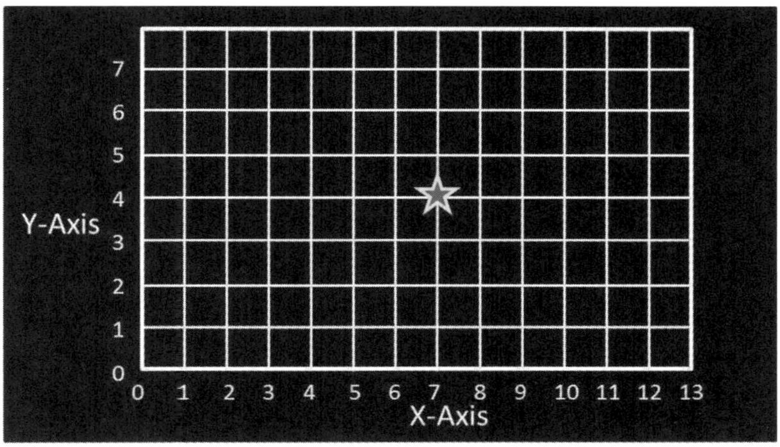

Fig. 4.1 Diagram of the basic X and Y machine motion determination.

Today's simulation software packages provide integrated tool, part, and machine motion analysis to accurately calculate the X and Y envelope.

The second step is to determine the position and range of the Z-axis machine motion. Z-axis machine motion determines how far back from the part the machine must sit to safely clear the part when in motion or at rest, and how far out from this position the machine must reach at its farthest required motion to perform the required operation.

4.2.2 Machine Types

A variety of machines take differing approaches to meet the process requirement. They range from simple to complex and from small, flexible systems to large monuments. "Monuments are tools, equipment, or machines designed for a specific purpose in automation. Once that purpose changes, these systems are very difficult and expensive to reconfigure to the new purpose, and therefore often become obsolete. Monumental Automation prevents easy reconfiguration of a production line upon new customer demands." [4.2]

Figure 4.2 is a "monument" type automated D/C machine. These large machines reflect the traditions of large machines with beds where parts are mounted. They require reinforcement of the flooring or a foundation. The process is to move the tool or product to the machine and perform the D/C operations. Typically, the assembly is then moved out of the machine's envelope to make way for the next assembly. The machine's cost and complexity makes it a valuable asset. The machine's rigid single-purpose design makes it inflexible. The monument type normally is a gantry machine. Figure 4.3 shows a unique underside drilling machine of a "monument" gantry configuration.

Fig. 4.2 "Monument" type automated D/C machine.

Some automated machines are independent of the work cell, as shown above, and others are highly integrated into the production process. Figure 4.4 shows a highly integrated assembly center incorporating loading, liquid shim, simultaneous two-sided D/C, and unloading station for a vertical stabilizer.

Fig. 4.3 Unique underside drilling machine of a "monument" gantry configuration.

Fig. 4.4 Highly integrated auto-drill/countersink assembly center.

This center uses a double-sided machine with an end effector on each side, with coordinated double-sided D/C. While drilling is taking place, the parts are loaded for D/C in the loading station. This type of integrated cell provides rigidity of the preassembled parts as it is transferred to the drill/countersinking position. Other transfer methods such as cranes can "tweak" structures and cause loss of positional integrity. Concern as a result of the build and lift-to-move (crane) approach results in reevaluation and validation of the structure's position after it is loaded into the drill/countersinking cell.

Another type of machine is a mobile machine that attaches and detaches from assemblies. The machines are moved to and from smaller assembly tools and attached for performing the D/C operation. They are moved into position by electric motor, pneumatic propulsion (air bearings), or crane. Precision attachments align the machine to the tool. Figure 4.5 shows the Mobile Gantry Applied Drilling System (MOGADS) invented and engineered by Frank Chang of Northrop Grumman Corporation [4.3].

Fig. 4.5 Mobile Gantry Applied Drilling System.

MOGADS provides D/C to a group of eight fixtures containing assemblies. The advantage of the system is its flexibility and ability to go where it is needed as fixtures are loaded. The MOGADS type design is popular with cost center managers due to its use-where-ready capability. MOGADS is part of the highly successful line of Gantry Applied Drilling Systems (GADS) employed by Northrop Grumman Corporation on the assembly line in El Segundo to enhance quality and reduce cost.

Mobile drilling technology has reached into the realm of exotic with the application of crawling robotic driller robots on Airbus structures. Figure 4.6 shows the highly mobile crawling auto-driller, the Torreslightdrill robot. The machine weighs in at 47 kg (103 lb).

The Torreslightdrill robot uses a laser system to establish its position in relationship to the part. Suction cups lock the robot into position where it drills and countersinks a group of holes. Once the robot has completed the drill/countersink operation, the laser guides it as it walks to the next group of holes to be drilled and countersunk.

Fig. 4.6 Torreslightdrill robot.

The desire for lower-cost lightweight and highly flexible systems has driven new auto-drill systems. They have been enabled by the increased ability of industrial robots used primarily in the automotive industry to meet aerospace positional hole tolerances. The United States Government and industry funded research to develop a drill countersinking system to access and drill/countersink difficult areas for humans or traditional auto-drill systems to access. Figure 4.7 shows the concept of operations for an auto-drill/countersink system to operate inside an inlet duct.

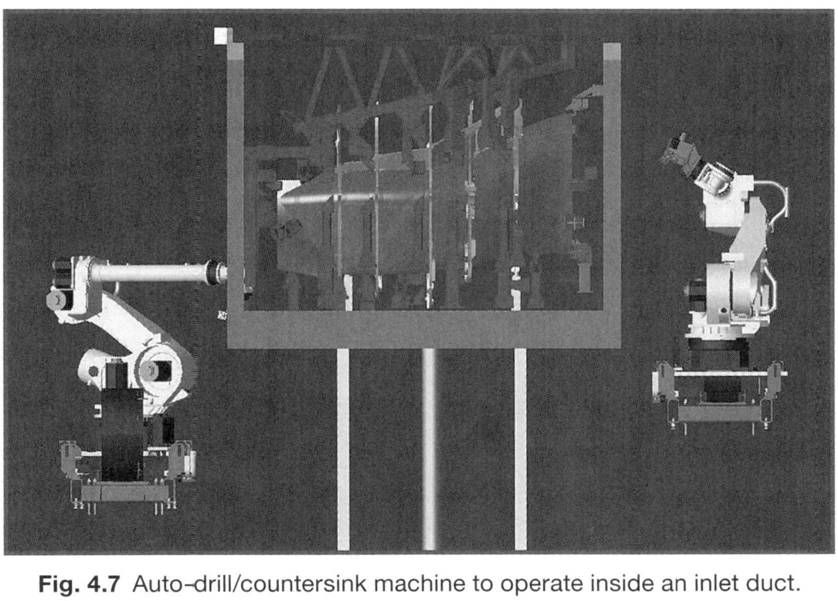

Fig. 4.7 Auto-drill/countersink machine to operate inside an inlet duct.

The machine's small end effector and laser positioning and confirmation systems were major breakthroughs in the use of articulating arm robots for precision aerospace D/C in confined spaces. The inlet automated D/C operation is part of the Integrated Assembly Line (IAL) located in Palmdale used to produce the F-35 center fuselage.

The line uses robotics automation to increase affordability, quality, and efficiency, according to company officials [4.4].

Not all the examples of auto D/C machinery must be complex, large, flexible, or exotic. The best application of automation is the simplest that can be applied to meet the requirements of the process.

Figure 4.8 shows two small auto-drill machines mounted on assembly fixture.

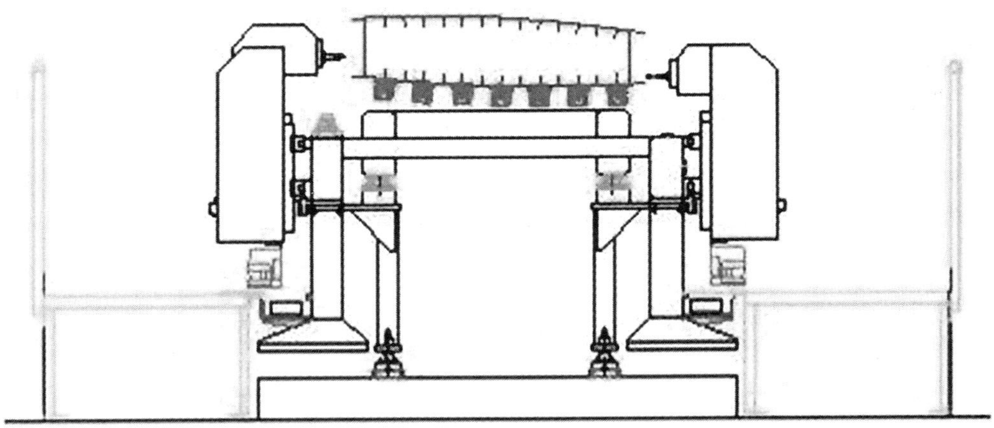

Fig. 4.8 Two auto-drill machines mounted on assembly fixture.

They have one horizontal axis and one short vertical axis. A movable spindle extends to drill the hole once the correct position is reached and confirmed. These machines are being used for drilling and riveting wing panels and spars.

4.2.3 Tools and Parts

Parts are combined to form assemblies. They are held in position by tools while holes are drilled and countersunk. Fasteners of differing types are installed into the holes. The assembly then becomes a stand-alone component of the airplane. It can be placed aside until it is joined with the next sub-assembly.

When the components are drilled, countersunk, and fastened by hand, the mechanic moves in and around the subassembly in a continuous process of drilling, countersinking, and validation. Validation is performed by the mechanic because the hand process is subject to variation. Each subassembly's components represent a new set of positions different from the previous assemblage.

When tools and parts are married together for automation, new tooling approaches have to be developed to facilitate the uninterrupted operation of the machine. To do this, the tooling must be evaluated, designed, or redesigned to enable precise repeatability of the component parts of the subassembly.

The first step to remove variability is to unmask the areas that contribute to instability. Once these hot spots are mapped, the tools can be modified to attain the stability needed for automated drilling.

Tooling designed for hand operations does not account for the effects of a machine that weighs thousands of pounds moving in and around its structure. Deflection can alter component positions. Therefore, a secondary evaluation to unmask variation derived from machine motion should be performed and anomalies corrected.

4.2.4 Integration Level I

Assembly tools, parts, and the machine are combined for one purpose: to accurately perform the D/C operation. Holes are drilled to exacting tolerances—0.000 in to +0.0015 in (−0.000 mm to ~0.4 mm). Therefore, any slip in the collected assembly during the drilling process of hundreds if not thousands of holes will result in the misalignment of holes that will prevent the installation of fasteners.

When automating the D/C process, several methods are used to integrate tool, parts, and machine. Included are moving the clamped-together assembly to the machine; moving the tool that holds the parts in position and the clamped-together assembly to the machine; or assembling the components of the assembly in front of the machine and never moving it until it is drilled. The option to load the assembly in front of the machine is the same as the hand method. The problem arises because of the expense of the machine and the need to optimize its utilization. Many hand workers can migrate between rate tools. But there is only one machine station. Therefore, the best second option is to have mobile assembly tools that are constructed with the rigidity necessary to maintain the integrity of the location of the components.

4.3 Action

Once the machine motion is established, an action must be performed to accomplish the D/C operations. The actions necessary to carry out all the operations associated with D/C of a hole are performed by an end effector.

4.3.1 End Effector

The end effector performs a complex set of actions that are coordinated to successfully D/C of a hole. An end effector's functions must be packaged small enough to access holes in and around structures and to minimize weight that adds to deflection at the tool tip. Actions that the end effector performs include holding the tool that drills and countersinks, vacuuming to collect dust, use of sensors to determine over-limits, scales to provide feedback, a pressure-foot to compress the work-piece, and a coolant distribution system. Figure 4.9 shows an end effector used for D/C holes on airframe assemblies.

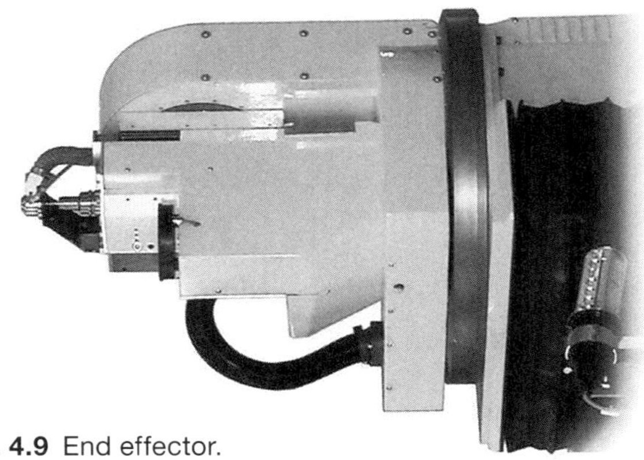

Fig. 4.9 End effector.

End effectors are highly specialized pieces of hardware that are either produced by the machine integrator or purchased separately from companies that specialize in their design and manufacture. If purchased separately, the end effector is provided to a machine integrator who attaches the end effector to the machine.

4.3.1.1 Size

The access required to acquire as many holes as possible places design pressures on end effector engineers. In many cases, the end effector must be designed small enough to fit into tight spaces. In addition, minimizing size also provides more flexibility and freedom when moving in and around parts, tooling, and obstructions. The design requires size minimization while including all the attributes necessary to carry out the requirements of D/C. Included in the design requirements are the hole evaluation requirements to maintain continuous speeds and controlled feeds through the material designated for the end effector.

4.3.1.2 Power

During the speeds and feeds regime, the end effector must perform high speed to low speed and high torque operations. While minimizing size, the design engineer must also provide a strong enough motor to supply continuous power and torque to the spindle for the stacked materials it drills and countersinks. This includes high speeds and feeds for drilling through composites and low speeds with high torque when used for drilling through titanium and steel. The positional accuracy and diameter requirements for holes drive the requirement for a motor that can sustain operations without degrading performance or accuracy due to heat.

4.3.1.3 Capability

The capability of the end effector includes a quill that feeds out and is integrated with a pressure foot attached to a measurement device that provides feedback to the controller. As the quill is fed out, the pressure foot advances and touches the structure. Upon touching the structure, the scale reader is compressed, which alerts the controller to spin the motor and begin the drill countersink routine for that position. The scale continues to send signals to the controller as it progresses through the material stacks, changing speed and feed rates as different materials are encountered. In addition, the scale feedback mechanism alerts the controller to activate a pecking routine to clear chips. A pecking routine drills through material to a specified depth and rapidly retracts so that the drill is clear of the hole to clear chips as the result of drilling. Once the chip is cleared, the drill reenters the hole and begins drilling at the point that it left off before retraction.

While not as glamorous as all the other capabilities of the end effector, it also includes a lighting component for illuminating the work-piece during operation.

4.3.1.4 Pressure Foot

The pressure foot is the finger that touches the surface of the part. The pressure foot performs several functions. It is a vibration-absorbing mechanism of an automated D/C machine: having a shock to mitigate communications of vibrations to the drill tool, while simultaneously dampening the vibrations at their source to increase control of the drill tool during D/C operations. The pressure foot of the end effector is upwardly rotatable to access the drill tool to the chuck [4.5].

The pressure foot also has a pneumatically or hydraulically driven pressure cylinder to facilitate pressure against the part and substructure to minimize separation.

Figure 4.10 shows an end effector engaging the structure of an airframe. The end effector must be designed to allow airflow in and around the spindle for chip evacuation. The end effector, when installed, is a precision removable (for maintenance) part of the machine where its length is coordinated with the scale feedback system.

4.3.1.5 Tool Holder

Tool holders come in different configurations and sizes. Precision tool holders are used to clamp the tool and countersink into the spindle of the end effector. They are inserted into the machine, and their size is dependent on the size of the end effector and the size of the drill they will hold. The tool holders are preset in the tool setup cribs to a precise depth to ensure the correct activation of speeds and feeds as the machine progresses through the material stacks. The drill tip is set in relationship to the surface of the pressure foot. The countersink is also set in relationship to the pressure foot surface. This is done to ensure that the machine "knows" when it has touched the surface of the product, precisely how far it is before the drill impacts the product.

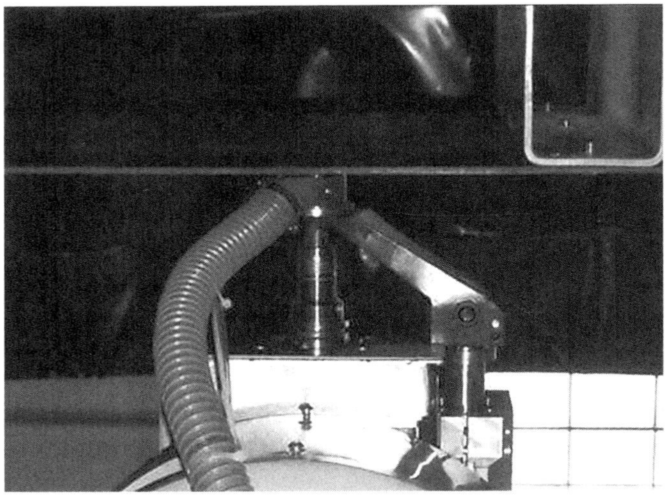

Fig. 4.10 End effector engaging the structure of an airframe.

4.3.2 Dust Collection

Composite materials have spread across the aerospace industry as the material of choice, especially for external skins. The process of D/C composites creates dust particles that can spread across the factory. As a result, the integration of specialized vacuum systems into the end effector and coordinated with its operation has become part of the design criteria. Figure 4.11 shows the centrifuge dust collection system in which dust particles are collected from the vacuum system of an automated D/C end effector.

Fig. 4.11 Centrifuge dust collection system.

4.3.3 Coolant

Coolant for D/C is delivered through the end effector to the tool tip as mist. The integration of mist coolant is necessary, especially where substructure materials are harder than the composite skins.

4.3.4 Tool Changer

Tool changers eliminate the need for human interaction with the machine to change drills and countersinks for the next hole size or to replace worn or damaged drills. As a result, the end effector must have the capability to access the tool holder inside the pressure foot. The automated nature of the tool change has to include a tool change routine that lifts the pressure foot, releases and stores the used drill, accesses the new drill, acquires the new drill and clamps it into position, and then closes the pressure foot.

Modern tool changers have accessories that validate the D/C set length as an added safety measure before sending the machine off to D/C more holes.

4.3.5 Vision System

Vision systems are multifunctional: their use may be passive or active.

Passive vision systems serve as operator observation media integrated into the Human Machine Interface (HMI). The image and observation of the machine's operation provides the ability of the operator to stop the machine if they visually detect a suspicious event or anomaly.

Active vision systems can provide a level of fidelity for quality evaluation of the substructure position or hole/countersink inspection in process. The vision system quality component has had mixed success. Vision systems are traditionally subject to lighting and debris that affect their precision and performance.

Early in the development and initial transition to production of automated D/C machines, effort was made to use vision systems to replicate the hand drilling method. The plan was to use the vision system to evaluate the substructure edge and compare it to the drill routine to see if the holes would conform to the specified ED. If needed, the holes that would not be in conformance would be shifted to a position to meet ED specification. This approach was rejected for several reasons. One was the time needed to evaluate the substructure with the machine. The other was the level of complexity that the software would need to compare, contrast, and alter the drilling NC program.

4.3.6 Inspection

Integration of inspection devices to do in-process hole and countersink inspection has been the next technological evolution of the end effector. Current systems use a probe installed in place of the drill/countersink. An inspection program is run to confirm that quality meets specified limits. The problem with this method is machine downtime and non-value-added work. While the machine is running an inspection program, it cannot drill/countersink holes. The opportunity lost is minimized during low-rate initial production but becomes more critical as the machine's utilization reaches 100%.

4.3.7 Integration Level II

Integration of all this activity must be tightly coordinated to minimize events that can cause damage to the part and the machine, meet quality specifications, and reach maximum efficiency. The integration of the machine to stop at position (point vector and normal to surface), lock in place, reach out to the structure, and activate the drill/ countersinking routine is the result of a complex interaction. Interaction and coordination takes place between electromechanical, pneumatic, hydraulic, and optical hardware. The interaction and coordination is driven by software programmed from engineering Computer Aided Design (CAD) and Computer Aided Manufacturing (CAM) systems.

The programs are downloaded into the automated D/C controller in their machine command format.

4.4 Controller

The controller is the medium between the human and the machine. It is the method of communication that permits the human in the loop to activate the machine and cause it to perform a preprogrammed task. It also allows the human to perform certain on-demand tasks outside the preprogrammed requirement of the machine by authorized and trained personnel for very specific activities. These activities include maintenance, calibration, clearing the machine away from structures, and certain tests.

Controllers can be highly complex and expensive units such as a Siemens 840D controller, or as simple as a single-purpose PLC programmed from a laptop.

4.4.1 Human Machine Interface

The HMI is the medium that communicates information for input, output of performance, alerts, alarms, notifications, and progress to the operator of the machine. The detail and face of the HMI is determined by the requirements document and its specified degree of fidelity required, causing the machine to operate. The HMI also has input capability for human operators to engage and command the machine to operate preprogrammed routines as well as command for specific motions.

Incorporated into advanced HMIs are interactive multimedia training systems that incorporate text, audio, and video graphics animation and other tools to teach an operator to use the automation or other tool. The training system includes a computer workstation to allow inputs from an operator, computer memory, and a training computer program operating on the computer that has a graphical user interface for interfacing the operator with the workstation and interactively training the operator to operate and perform tasks on the machine tool. The operator can provide inputs to the training system by various input devices such as a keyboard, cursor, or mouse. The training system additionally includes interface software operating on the computer for controlling the machine tool and sharing access to and exchanging data with the computer memory and a remote station linked to the computer memory for allowing an instructor to monitor and supervise the operator in real time [4.6].

4.4.2 Volumetric Compensation Algorithm

Certain software that enhances the performance of the machine to meet a greater degree of accuracy may be resident in the controller. Compensation programs have been used to correct the errors in the mechanical properties of the lead screws and gears that drive the machine to positions commanded by the software. Scales help coordinate and correct the mechanical variations. They do perform the corrections on the linear and on some machines on the rotary axis. The combined accuracy corrections are at the point of the motion and not at the tool tip. The accuracy needed for automated D/C machines is at the tip of the drill. Highly specialized software compensates for mechanically derived error by correcting the error at the drill tip.

Another type of compensation system uses a laser feedback system from the drill tip to measure position and correct any missed positions before the drill routine begins.

4.4.3 Integration Level III

Integration of all of the machine motion, machine action, position, and interface with the tooling and product is very sophisticated in its most complex form. But there are simple applications where the coordinated actions of the machine do not require higher-level systems, programs, operation, or support. All of these systems need some level of coordination and integration to ensure their proper use when they touch an assembly. The three separate components of the process, tools, parts, and machines need to be aligned and coordinated, no matter how small or simple the machine or assembly.

Humans can observe a static display of parts positioned within a tool and place their drill bit where necessary to obtain a desirable outcome. Machines assume that the parts are placed in the proper position and move to a preprogrammed position and begin operating. Defects will occur without the integration of tools, parts, and machine.

The difference between a machine tool for fabricating parts and a machine tool or piece of automation for D/C on-assembly is large. A machine that makes single parts that creates a defect and scraps a part is recoverable. Of course, even a scrap defect on an assembly is recoverable. But scrapping a part usually occurs early in the process when the investment cost is low. Scrapping an assembly or part of an assembly is very expensive because so much has already been invested in the collected parts. Most if not all of them have been through all of the processes except for their joining together. That is why the integration of all the components of the on-assembly automated D/C system is important.

4.5 Calibration

After all the work is performed to integrate the various operating components of the D/C machine, the calibration is performed. Initial and continued validation and calibration is required to ensure that the parts, tooling, and machine act in harmony. Figure 4.12 shows one of the precision measurement systems used to calibrate and certify accuracy and coordination between machine, tooling, and parts.

Fig. 4.12 Precision measurement system.

The laser tracker is set up in the assembly drilling cell after the parts are loaded into the tool. The machine is tracked over its entire hole-to-hole motion to confirm its ability to meet specified operational specification. The laser tracker also checks the substructure position of the parts to ensure that the drill will penetrate and result in the proper edge distance.

Laser trackers are used for dimensional certification because of their high accuracy, portability, and reliability. These devices provide the combination of compact size, high accuracy, and software support to integrate positional measurement into automation processes.

The current laser tracker is a point-to-point tracking system that requires intrusion into the operating space of the assembly cell. The intrusion interrupts the operations in the cell while the tracker measurement operation takes place. Each hole and motion must be tracked as a separate event. The point-to-point tracking method consumes valuable time that the machine could have used to D/C hole.

A number of new methods have been developed and are coming online to offer rapid assessment and measurement for an entire surface or machine envelope. Two of these systems are laser radar and holographic imagery. Both types of system map a surface or motion and are less intrusive into the workspace, freeing the machine to do what it was built to do, D/C holes.

4.6 Sustainability

One of the important considerations for any of the machines a company chooses is sustainability. The cost and effort to select a machine should include the spares and equipment necessary to keep the machine in operation. Operational sustainability cost is driven by the complexity of the machine selected. Operational sustainability is also complicated by the sophistication, size, and complexity of the machine selected. The short view is to purchase a machine without consideration of all the critical elements for its continued operational health maintenance throughout its life cycle. This includes the assessment of the long-term ability of the critical parts supply base to sustain the operations and support over time.

4.7 Summary

The selection of the appropriate piece of automation begins with the requirement that the machine will accomplish. The determination of size to reach all the holes defined in the requirements document is the first step in the process. There are many types of machines that have the capability to meet a wide variety of size requirements. Some are "monument" types while other offer mobility and still others provide flexibility.

The integration of tooling, parts, and the machine is more complex than the interaction between tools, parts, and humans. When automation is applied, the tooling and part relationship is evaluated to ensure that the motion of the machine is infused into the process successfully.

The end effector is the business end of the automated D/C machine. It contains all of the technology that enables the on-assembly D/C of holes.

The integration of the motion of the machine and the end effector is a highly complex operation using a variety of technologies acting together. The motion of the machine and the action of the end effector is controlled and coordinated through the software and NC program.

The controller also uses sophisticated software to enhance the precision of the machine.

All the actions of the machine are calibrated and confirmed using laser tracking machines to ensure that the machine, tooling, and parts are aligned for successful D/C.

A variety of machine types are available in the marketplace to evaluate and determine if they meet the needs of a company's on-assembly drill/countersinking requirements. It is easy to buy capability where it is not needed. Requirements creep and attraction to unneeded capability can inflate the cost of the machine and complicate integration of the machine into a production line.

Simplicity is best. Further capability can be added later when the machine has demonstrated its basic purpose successfully: D/C of holes.

Automation should be applied only where it makes sense and where several of the following will result: higher output, better quality, reduced scrap and rework, improvements in workspace safety, and fewer required people.

References

4.1 Storaasli, O., J. Ransom, and R.E. Fulton, "Structural Dynamic Analysis on a Parallel Computer: The Finite Element Machine," *Computers and Structures*, 1987, Vol. 26, No. 4, pp. 551–559.

4.2 Kihlman, Henrik, "Affordable Automation for Airframe Assembly, Development of Key Enabling Technologies," Linköping Studies in Science and Technology, Thesis No. 953, 2005, p. 11.

4.3 Chang, Frank and George Bullen, Mobile Gantry Applied Drilling System, U.S. Patent 5836068.

4.4 "First F-35 Center Fuselage Delivered From New Assembly Line," airforce-magazine.online, online Journal of the Air Force Association, April 03, 2012, http://www.airforce-magazine.com/DRArchive/Pages/2012/April%202012/April%2003%202012/FirstF-35CenterFuselageDeliveredFromNewAssemblyLine.aspx, Retrieved June 2, 2012.

4.5 Bullen, George N., Vibration-absorbing end effector of an automated drill and countersink machine, U.S. Patent number: 6231280, Filing date: Aug 10, 1991, Issue date: May 15, 2001.

4.6 Bullen, George N., Machining tool operator training system, U.S. patent number: 6033226, Filing date: May 15, 1997, Issue date: Mar. 7, 2000.

Chapter 5
Considerations Before Replacing Hand Drilling with Mechanization or Automation

All change is not growth, as all movement is not forward.

—Ellen Glasgow

5.1 Introduction

The first question that a person, organization, or company must ask is, "Why do I want to automate this process?"

The obvious answer is that there are benefits such as lower per hole cost, increased quality, consistent throughput, and fewer repetitive motion injuries. When added together they make a strong case for automating D/C of every hole on the airframe.

This chapter will address the considerations necessary for automation application where a successful first application of automated D/C has been installed and where it has never been applied for D/C holes.

Within the factories that have successfully applied automation, the euphoria of success tends to overwhelm common sense and provide inertia to those that would automate the D/C of every hole on the airframe. The outcomes from a single successful application make converts of those questioning the validity of claims for positive returns prior to installation. The new zealots for automation reside in leadership, customers, and decision makers who grasp onto the first results as evidence that automated D/C will improve every other area where it is applied.

This has led to the use of data derived from tools such as ROI calculations that provide singular justification for expansion of the technology. Direct cost savings alone is often used as a justification for adding machinery without consideration of ancillary effects to the operation of the factory. This leads to the rapid expansion of automated machinery without judicious consideration for issues such access to the holes to be drilled and countersunk, infrastructure, labor, and facilities. The rapid expansion can and has overwhelmed the ability of the operation to absorb the change and integrate multiple installations into the production stream simultaneously. Difficulties and failures result that sour the taste for continued automation. Therefore, other decision tools have to be a part of the first installation or expanded application of automated D/C machines. They include a feasibility analysis, cost benefits analysis, and risk assessment.

The first application of automation usually incorporates a detailed master plan for installation on the airframe where that greatest chance of success will be realized. The cost center where this section resides includes all the attributes for the best opportunity for success. This does not mean that every other cost center in the factory contains the same "golden nugget" of opportunity. Decision models have been developed to ascertain the benefits derived from applying automation, starting with the best to least viable. They take into consideration every aspect of the factory's sensitivity to change including the rate that the factory can integrate automated D/C into the factory operations. These models and decision matrix templates are provided as part of this chapter.

A group of proponents in aerospace supports the philosophy that every hole on the airframe should be drilled by automated machinery. This group resides in the sales department for automated machinery, controllers, and software. The proponents of the universal automated D/C provide a wide range of tools to develop positive cost, quality, and schedule data to support their claim. They target aerospace conferences and present claims of success to attendees who may be considering automation as an alternative to the hand D/C process. While aerospace conferences provide exposure to a wide range of options for consideration of automated D/C, they are used as a promotional medium.

Therefore, the tools contained in this chapter such as decision matrices will provide templates containing valuable objective considerations to offset the information provided to the attendees at these conferences. The provided decision matrices were derived from detailed trades and analysis. They validate the optimal process and provide risk assessment decision criteria.

Sometimes the best option is to leave the process alone.

5.2 Before the Decision to Automate

Once the decision has been made to consider automation as a potential replacement of the hand method for D/C holes, several evaluations must be made. The evaluations include a feasibility analysis and a cost benefits analysis. The feasibility analysis is required to determine the ability of the factory and tooling to facilitate the change to automated D/C. Assessment of each of the assembly centers, called *cost centers*, has to be performed to determine which, if any, are suitable to integrate automation in the cost center's operation. Once the feasibility analysis is finished, further analysis is performed to rank the candidate cost centers according to their optimal potential for integrating automation. The ranking is combined with the results of the cost benefits analysis to determine which cost center has the best chance of success and the greatest ROI.

5.2.1 Feasibility Analysis

The purpose of a feasibility analysis is to study and analyze the tasks necessary to determine the capability of the cost center, infrastructure, tooling, part fabrication processes, and factory facilities to accept the insertion of automated D/C into the cost center. The feasibility study is performed on an existing assembly line at any stage of its maturity or on a new assembly line in the design stage. The study is performed by a manufacturing engineer in small factories, or a team composed of a manufacturing engineer, industrial engineer, facility engineer, and manufacturing development engineer in larger factories.

Some of the primary considerations when performing a feasibility study on an existing assembly line are access by the machine to the holes to be drilled (obstruction), ability of the flooring to support the added weight of the machine, ability of the tooling to integrate with the machine without affecting the tooling integrity, factory infrastructure necessary to support the operation of the machine, operations within the cost center that interact with the automation (performance and safety), environment (personnel capabilities' assessment and space), labor, supplier ROM, and deliverable.

It should be noted that prior to beginning a feasibility study, the Social Concept of Operations (S-CONOPS) steps and considerations described in Chapter 12, Social Impact of Automation, are important to facilitate cooperation at all levels of the production operations organization.

5.2.1.1 Access

Access for the machine to reach the holes to be drilled is critical to maximizing the benefits necessary to justify the acquisition of a D/C machine. On an existing airframe assembly line, the tooling to hold structure in place limits access for the machine to reach the structure and to move between holes. Therefore, there may be an assembly that has a large numbers of holes, but no opportunity because they are obscured by structure.

Another type of obstruction that can interfere with the operation and motion of the automated D/C machine includes the work-stands that surround the assembly tool. When assembly tools are built, they usually have access work-stands fabricated that are independent of the assembly tool and fixed to the floor. The work-stands provide mechanics with access to the product and the assembly tool. The feasibility study includes the design for reconfiguration or mobility to gain greater access to potential holes in an existing factory. In new factories, consideration must be given to the work-stand as an integral part of the machine, assembly tool, work-stand, mechanic access integrated design criteria. In many cases, the design or redesign of work-stands that are removable need to be incorporated for machine operation. The cost to modify existing work-stands to mobile work-stands is included as part of the cost benefits analysis.

To perform an access analysis, a generic machine size must be developed, including access from all the angles necessary for the drill/countersink tip to reach the structure. The analysis must also include machine motion between holes and the motion of the machine to exit the tooling and return to its home position. The best medium for this kind of generic machine dimension and motion construction is the use of CAD models and simulation. Figure 5.1 shows the orb-of-impact plus safety dimension and machine arm notional configuration.

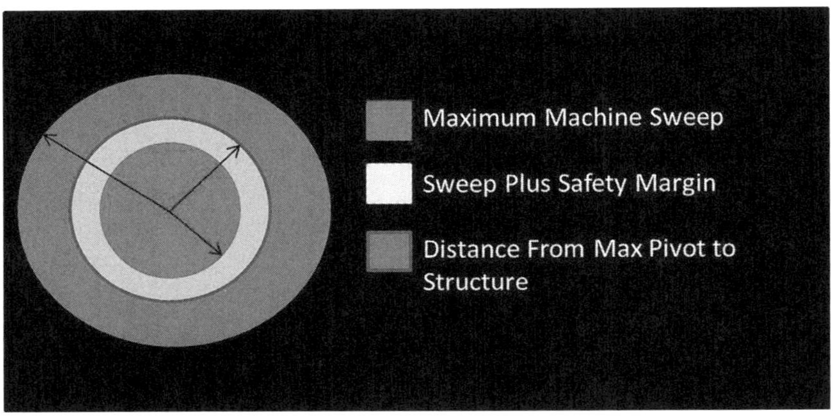

Fig. 5.1 The orb-of-impact plus safety dimension and machine arm notional configuration.

This type of access analysis can also be done effectively using laser tracker, optical equipment such as transits and levels, or even a tape measure.

Of course, the best way to visualize and manipulate the generic drill head and arm is through a robust CAD model. CAD-generated simulation models provide a greater degree of fidelity and also provide the ability to generate motion studies for machine optimization. The simulation models are a valuable tool to mitigate management concerns by conveying a visual representation of the machine's ability to traverse the assembly tool and product safely.

A matrix of results obtained from the feasibility study is included in the report and briefing to leadership. Figure 5.2 provides an example of what an access feasibility matrix should look like when populated with notional data.

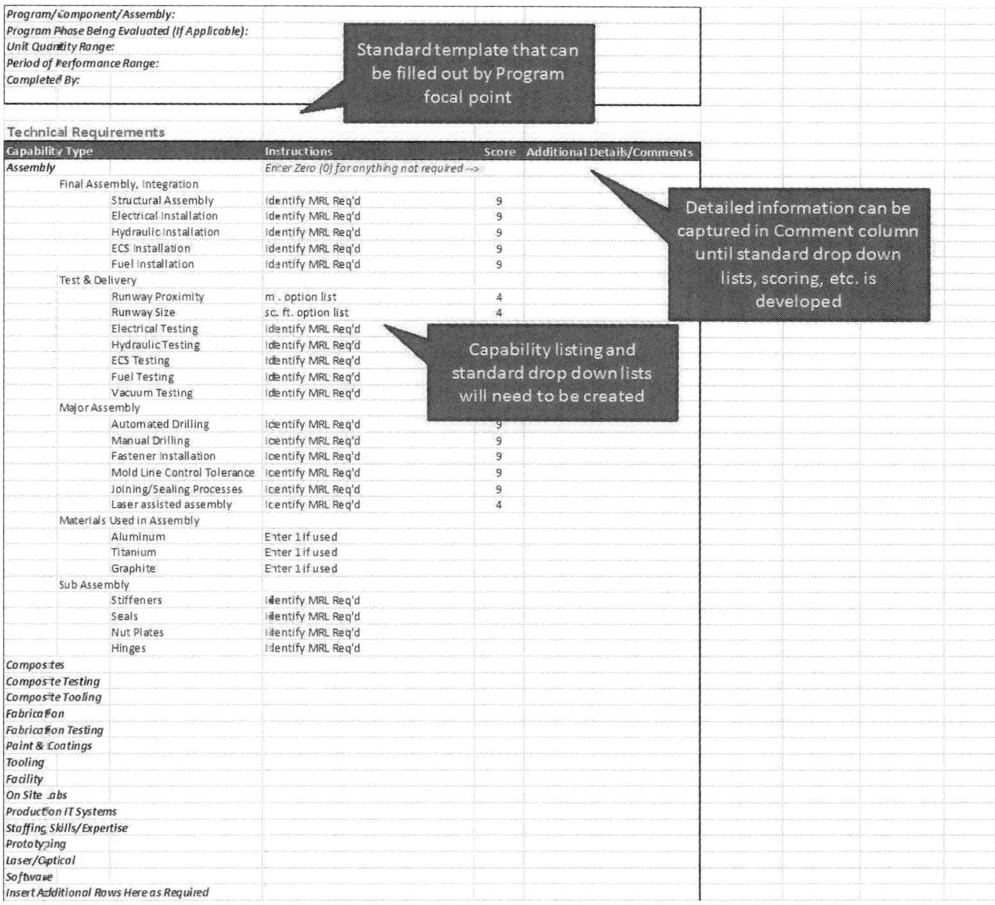
Program/Component/Assembly:
Program Phase Being Evaluated (If Applicable):
Unit Quantity Range:
Period of Performance Range:
Completed By:

Technical Requirements

Capability Type	Instructions	Score	Additional Details/Comments
Assembly	*Enter Zero (0) for anything not required -->*		
Final Assembly, Integration			
Structural Assembly	Identify MRL Req'd	9	
Electrical Installation	Identify MRL Req'd	9	
Hydraulic Installation	Identify MRL Req'd	9	
ECS Installation	Identify MRL Req'd	9	
Fuel Installation	Identify MRL Req'd	9	
Test & Delivery			
Runway Proximity	mi. option list	4	
Runway Size	sq. ft. option list	4	
Electrical Testing	Identify MRL Req'd		
Hydraulic Testing	Identify MRL Req'd		
ECS Testing	Identify MRL Req'd		
Fuel Testing	Identify MRL Req'd		
Vacuum Testing	Identify MRL Req'd		
Major Assembly			
Automated Drilling	Identify MRL Req'd	9	
Manual Drilling	Identify MRL Req'd	9	
Fastener Installation	Identify MRL Req'd	9	
Mold Line Control Tolerance	Identify MRL Req'd	9	
Joining/Sealing Processes	Identify MRL Req'd	9	
Laser assisted assembly	Identify MRL Req'd	4	
Materials Used in Assembly			
Aluminum	Enter 1 if used		
Titanium	Enter 1 if used		
Graphite	Enter 1 if used		
Sub Assembly			
Stiffeners	Identify MRL Req'd		
Seals	Identify MRL Req'd		
Nut Plates	Identify MRL Req'd		
Hinges	Identify MRL Req'd		
Composites			
Composite Testing			
Composite Tooling			
Fabrication			
Fabrication Testing			
Paint & Coatings			
Tooling			
Facility			
On Site Labs			
Production IT Systems			
Staffing Skills/Expertise			
Prototyping			
Laser/Optical			
Software			
Insert Additional Rows Here as Required			

Fig. 5.2 Example of what an access feasibility matrix should look like when populated with notional data.

5.2.1.2 Flooring

A concern when adding weight to flooring is the ability of the floor to support the added weight of the machine. Additional consideration for machine motion and shifting loads that can cause floor deflection or sag, which in turn can affect machine accuracy, must be evaluated for acceptability. The floor material type and grade, thickness, and soil compaction under the flooring must be analyzed from existing documents if available or from core samples.

The facilities engineer or a consulting group should perform the analysis, and the results should be incorporated into the feasibility analysis final report and briefing.

The review and evaluation of the flooring will result in a determination of the need to reinforce the floor to accept the added weight and motion strains. If needed, these modifications can interrupt the existing production operations and result in costs that are above the direct cost to reinforce the floor. The feasibility study and analysis includes the effect on production but also includes the opportunity options when floor reinforcement can be done with minimal effect on production operations. One example that can mitigate the production operations impact of floor modification for strengthening is annual shutdowns for assembly tool maintenance and calibration. Flooring modifications can be coordinated and performed during the shutdown to partially or wholly eliminate disruption to production operations.

5.2.1.3 Tooling Integration

Automated machine integration with airframe assembly tooling is compounded by the supporting tools that facilitate the mechanic's ability to position, hold, drill, countersink, and fasten the airframe assembly together. These include removable details, drill bonnets, and templates. The determination if, and to what extent, the assembly tooling and its operational support tools need to be modified or eliminated to accommodate the automated D/C machine is part of the feasibility study. How much tooling should be retained for "backup" to support a comfort level for production continuity if the automation schedule slips, or the machine gets canceled or fails to perform, is also part of the study and analysis.

Automated D/C machines are either attached to the assembly tool or stand independent of the assembly tool with the machine supported by its own structure. In either case, modifications are done to reinforce the assembly tool to accommodate the added weight and motion of the machine where the machine is attached to the tool; or reference systems are added to the assembly tool to align the machine to the assembly tool, and the components of the airframe are assembled within the assembly tool.

The feasibility study includes the options and benefits of each method with cost and trade studies for decision considerations.

The assembly tool and its supporting ancillary tooling are the core of the process for assuring that quality products meet specifications. Any change to the assembly tools and their ability to hold the various internal and external components of an airframe assembly in a fixed position throughout the fastening process must be thoroughly analyzed and evaluated for the effect of the change. This is especially true because the pressure exerted by the machine includes the pressure foot and the added pressure from one-shot drilling. The automated D/C machine will deflect the substructure and cause separation from the skin when the drill passes through the skin. The lower pressure from the hand drill method will not cause the same deflection because the holes are step drilled, which is accomplished with much lower force. Substructure deflection and snap-back cause chip buildup between the skin and substructure, elongated holes, chip damage to the skin, burrs, and other defects. Therefore, the feasibility study includes the deflection analysis and determines if, where, and how much reinforcement must be done to the assembly tool to accommodate the added forces from automated D/C.

5.2.1.4 Factory Infrastructure

Introduction of change into the factory of any piece of machinery that is different from the norm can cause failure. Factory operations have departments, equipment, processes, and suppliers that are dedicated, trained, and integrated into the day-to-day support for tools, equipment, and machines that make possible the continued uninterrupted assembly of product.

The feasibility study includes an assessment of the ability of the factory and its suppliers to expand their capability to meet the demands for support of the new machinery. Internally, support departments must be able to respond rapidly to requests from production to reengage a machine brought out of action by damage, malfunction, or unresponsiveness. The investigation of capability includes NC Programming, NC machine maintenance, personnel capability and competency, spares facilities, calibration capabilities, and calibration equipment types.

Externally, the geographic response distance (calculated by emergency response time from call to site) of qualified maintenance personnel on contract to respond to an emergency is determined. The installation of custom-built automated D/C machinery requires specialized integrators that are less than 24 hours from call-to-site (CTS) for rapid return of the asset to production operation. Outside the 24-hour time CTS, the feasibility study provides production impact analysis data to form the acceptable decision model. The feasibility study provides the decision information as impact data to help leadership decide where acceptability levels for production downtimes have been exceeded.

5.2.1.5 Current Operations

Very few new airframe production lines are being built. Therefore, with few exceptions, most of the automated D/C machines being considered for integration will be into existing assembly lines.

Existing assembly lines require a complete assessment and evaluation of the current operations to determine the level of disruption that the introduction of automation would cause. Hand operations have many different tasks including D/C that occur in close proximity to one another simultaneously. The degree to which the introduction of automated D/C interrupts or prohibits the simultaneous hand activities must be evaluated to determine the disruption to the line flow and labor-task redistribution effects. The impact to the entire assembly cell is part of a comprehensive impact analysis that is included in the feasibility study. The information derived from the impact study is combined with simulation studies to determine if workforce/task realignment analysis results can compensate for part or all of the disruption. The impact study includes the cost of workforce/task realignment to accommodate the introduction of automation as well as realignment recommendations to minimize or eliminate the impact.

Two other areas incorporated into a feasibility study that will affect current operations are standard tooling (drills and countersinks) and the cribs that supply them; and the effect on fastener tolerance, supply, and installation. Automation complicates the supply of standard drills, countersinks, and fasteners. While in the hand method, drills and countersinks are supplied to mechanics that do self-setup and installation (for the most part) in drill motors, automation uses drills and countersinks that are precisely preset in their tool holders. Automated D/C machines require the precision to ensure performance. In addition, they are delivered to the automation in a kit that contains the entire run of drills and countersinks that will be used in the run.

Rivets that were supplied to the assembly line and that are installed into holes drilled by hand are built to a tolerance that has an allowable variance greater than can be installed when the same hole is drilled by hand. Therefore, the tighter and more consistent hole requires a more consistent fastener size with lower variance to the specified diameter.

The feasibility study incorporates the challenges and cost of change to bring the D/C crib up to the standards and incorporating the equipment necessary for automated D/C operations.

The feasibility study also incorporates the fastener supplier assessment to determine the supplier's ability to improve their performance and reduce the tolerance limits of the fasteners; and to measure, prepackage, and identify fastener sizes within runs. This is done to provide data to the drill procurement group and set up the crib to align the drill size with the fastener size.

5.2.1.6 Environment

Many of today's modern airframes are built in environments that are not conducive to the successful installation and operation of automated D/C machinery. The feasibility assessment includes the atmospheric conditions of the factory such as measurements of temperature and humidity variation, and vibration.

While these considerations may seem trivial on their surface, they have caused disruption, defects, and added cost to the operation of automated D/C machines.

Temperature variation in some factories can be 50°F (10°C). While the steel construction of the machine confines its coefficient of thermal expansion (CTE) for this temperature variation, the collective materials composing the skin(s) and substructure of the product and the tooling that holds it in position can expand and contract to defeat the specified positional tolerance of the drilled hole. While the machine may be stable throughout the temperature range of the day's drilling, the tooling and components of the assembly are moving in front of it.

Extremes and variation in humidity can cause lubrication issues, rust, and corrosion that increase the frequency of servicing and maintenance intervals.

Vibration, while rare, is not unheard of. In one case in Florida, the automated D/C machine was fitted with a seismograph integrated to a trip switch that paused the machine when vibration exceeded acceptable limits. The vibration from twice-daily trains that passed by the factory shook the factory and the machine, causing defects until the seismographic/switch system was installed.

An evaluation to determine the possible effect on the product and cost to bring the factory up to standard for automation are part of a feasibility study.

5.2.1.7 Labor

The economic condition and emphasis on jobs in the United States and Europe has created sensitivity to the introduction of any technology that reduces the touch labor for a task and therefore the number of people used to perform the task. The cost benefits analysis that follows uses as a baseline the number of hours that are reduced to perform a given task. The way automation reduces the cost of a task and by extension the cost of the product is the reduction in the cost for human labor.

As a result, the modern feasibility study includes an assessment of the readiness of labor to accept the automation as a replacement for human labor. The feasibility study has to include a thorough review of existing labor contracts and the sensitivity of labor to workforce reduction due to automation.

5.2.1.8 Risk Assessment

The feasibility study collects data that populates a risk cube and a risk reduction plan. The primary candidates that emerge as the most viable are stacked according to their risk, from least risky to most risky. The least-risk candidates are then isolated into individual risk reduction plans that identify areas of concern that need to be addressed and the actions required for mitigation of the risk, estimated time to reduce risk, and cost for risk reduction efforts. Figure 5.3 provides an example of a risk reduction chart tied to the risk reduction plan.

The deliverable for the feasibility analysis is a report that compiles all the elements of the analysis inclusive of the risk analysis and risk reduction plan. It also contains an executive summary and a summary statement with recommendation for continuation to the cost benefits analysis step and a recommendation that narrows the candidates to those with the lowest risk.

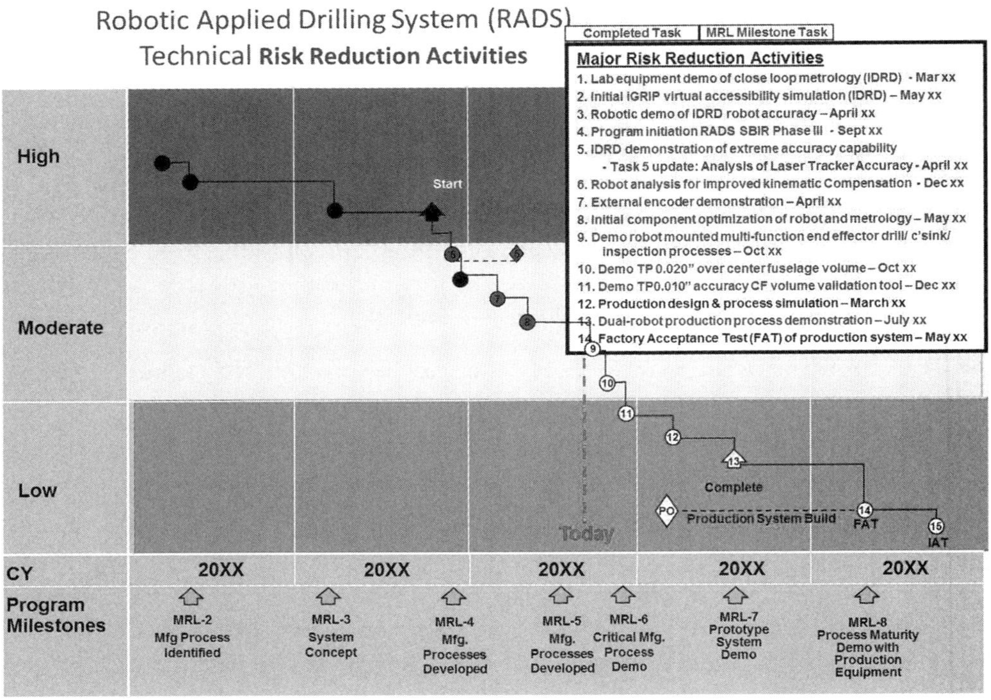

Fig. 5.3 A risk reduction chart tied to the risk reduction plan.

5.2.1.9 Feasibility Analysis Deliverable

The deliverable of the feasibility analysis is a document that contains data, information, and recommendations to facilitate the process for continuation for a go/no-go decision by the company leadership for installing automated D/C technology. As part of the study and analysis, the feasibility analysis individual or team provides the study and out-brief that contains a recommendation to proceed or not to proceed to the cost benefits analysis step. The study and out-brief are supplied to leadership.

5.2.2 Cost Benefits Analysis

If leadership decides to proceed to the cost benefits analysis step, the feasibility analysis document becomes a part of the cost benefits analysis. The cost benefits analysis individual team then begins to study the cost centers where the top candidates for automated D/C have been identified by the feasibility analysis recommendation.

The cost benefits analysis studies the existing hand D/C process for the candidate holes for automation. The study identifies touch labor hours required to prepare and drill the holes. It also collects estimated costs for drilling the same holes by automation, estimated machine costs for supplier provided Rough Order of

Magnitude Estimates (ROM-E), and combines the costs included for the cost center in the feasibility study to estimate a positive or negative cost delta to the hand D/C cost. The analysis also provides an ROI that determines the years to pay back the capital investment.

5.2.2.1 Hand Process Time Study Analysis

A time study of the hand drill countersink process is necessary because pure per-hole data does not exist for the specific holes that will be drilled and countersunk by automation. The selected holes that are included in the feasibility study need to be timed to collect touch labor costs directly attributable to the preparation and D/C of the specific holes proposed for automation. The time study is performed over several ship-sets so that an average time can be validated to D/C of the specified holes.

Observation of the hand operation is also necessary because historic performance measurements taken and reported within the cost center for D/C by mechanics and the cost center manager are suspect. They are suspect because the method used to collect the data varies and the activity within the cost center is centered on the collective performance within the cost center and not on the individual tasks performed. The data and reporting method for each task is minimized in the hand method in subservience to the cost, quality, and schedule output from the cost center of the product. In addition, any cost data for individual tasks are often hand written, difficult to obtain, and even more difficult to decipher and read.

The time included is all the time needed to drill the holes by hand, including setup time to prepare the D/C, insert the drill or countersink in the drill motor, retrieve and apply drill templates or drill bonnets, layout holes if needed, remove and store the drill templates or drill bonnets, and any other activities the mechanic performs to complete the drill countersinking operation.

Industrial engineers are the usual candidates to perform the hand D/C process. The data collected is compiled in a data sheet and becomes the benchmark for analysis to determine the economic viability of automating the D/C process.

5.2.2.2 Automated Drill Time Study Analysis

The automated D/C time analysis is more difficult because the automated D/C machine does not exist to perform an observation study. However, data exists that can be derived from machines in industry that have mature production D/C rates for the hole size, material stacks, thicknesses, and setup times. A trade study that incorporates a matrix of the quantities of holes for the size, thickness, and material type can be developed from the trade study data. The trade study is performed simultaneously with the hand touch labor time study and the other steps in the cost benefits analysis through Section 5.2.2 of this chapter. The trade study will provide data that is accurate if it includes several sources that have experience with D/C of the selected hole types. All of the data need not come from a single source but can be a compilation from multiple sources when all the required data cannot be obtained from a single source. This does not invalidate the accuracy of the data. Data from similar quantities of holes should be obtained to maintain accuracy of the trade analysis because setup times are included and amortized over the number of holes drilled and countersunk. If lesser holes are used, the per-hole drill/countersink times and movement between hole times can be extrapolated to estimate the cost for a specific run of holes.

The trade study excludes contributions and estimates from machine suppliers to avoid the image of biased data.

5.2.2.3 Machine Cost

The machine cost included in the cost benefits analysis is obtained from ROM estimates for machine suppliers who have partnered in the effort to determine the best technical solution for automated D/C machine applications. The ROM estimates comprise one element of the cost to be collected for development of the ROI. The estimate has to include any and all costs associated with the acquisition, acceptance tests, installation, integration, spares, training, and maintenance contracts to support the machine operations.

The cost does not include drill bits, countersinks, tool holders, or kit boxes.

5.2.2.4 Ancillary Costs

Ancillary costs include the effects of automation on the cost center environment and the current hand manufacturing method that its introduction disrupts. It includes costs associated with facility preparation such as foundation or floor strengthening, assembly tool modifications to accommodate the automation, and transition to production test materials.

It also includes the realignment of hand tasks accomplished concurrently that have to be postponed until after the automation performs the D/C operation. Many individuals within the cost center perform concurrent tasks that are often dissimilar and side-by-side. When automation is placed into operation, safety concerns are implemented that prohibit activities in close proximity to the machine's operation. The costs associated with the effect on efficiency through the cost center as well as realignment of feeder cost centers and receiving cost centers upstream and downstream from the automation have to be identified and captured. The cost benefits analysis has to include all costs over and above current costs for the hand operation.

The costs benefits from automation use pure touch labor savings from the delta between the setup and drill/countersink tasks consumed from the hand operation and the automated machine operation. All machine and ancillary costs are included in the cost benefits analysis; only touch labor costs are used to compare the machine to the hand-performed operation. The other attributes and cost savings, including lost time injuries reduction and improved quality, are not included in the cost benefits analysis.

5.2.2.5 Cost Collection and Conversion

When all the cost contributors are collected, they are used to compare and contrast the cost of implementing automation to replace the hand D/C process.

Before the comparison can be done, the time measurements must be converted to dollars. The variability of labor rates throughout the United States and Europe requires acquisition of the specific labor rate for the factory considering automation for D/C. This can be difficult, as labor rates vary from company to company, and from factory to factory within a company. The hourly conversion rate to dollars or euros also can vary from department to department within the factory. Companies also look at labor rates as proprietary and highly competition sensitive. They are reluctant to release the cost data for fear it will migrate to their competitors. Compounding these difficulties to obtain hourly conversion rates to dollars or euros are agreement by the factory leadership as to what dollar or euro amount should be used for savings within the cost center. Usually monetary amounts are fully burdened with overhead, and using the amount inflates the savings (if there are savings). Leadership, especially the cost center manager, is always conservative when assigning an hourly rate to the savings. This is due to the fact that savings (if automation is installed) is derived as a direct result of budget reduction within their cost center. The budget performance measure is reduced by the amount of savings projected by the cost benefits analysis conversion of hours to dollars or euros. Therefore, any reduction of the dollar or euro amount to calculate savings reduces the loss of budget to the cost center. All of these contributors result in a per-hour amount significantly less than the fully burdened amount used for bids and proposals or factory operational cost projections. A factory operating with a fully burdened cost of S150.00 an hour will generally arrive at a savings per hour allocation of $50 to $60.

Military airframe producers are more sensitive to the conversion of the hourly to monetary savings number. Commercial airframe producer's savings from manufacturing enhancements are improvements to their margin. The savings can be retained or passed onto the customer to increase the company's competitive posture. Military airframe producers are obligated to pass savings onto the customer. The intense scrutiny by government agencies requires the constant investigation and reporting of cost and performance numbers. Military airframe producers are always assuring their customer that they are seeking the most efficient means of production to reduce cost and improve quality while simultaneously understating the savings to retain margin. Automated D/C installations are highly visible evidence that can be shown and regularly demonstrated to government customers as evidence of their commitment to cost and quality improvement. They also affect the margin/profitability of the military airframe producer.

Therefore, the sensitive nature of hourly to monetary conversion has to be agreed to by all levels of leadership prior to calculating potential savings before conducting a cost benefits analysis.

Agreement on a number allows the conversion of the hours to dollars and enables the calculation of a comparison between the current hand process and the machine process for D/C holes. The cost benefits analysis presents a single ship-set of savings, without regard for a learning curve. It assumes that the machine will drill at its mature efficiency for the purpose of the calculation, without averaging in the aggregate of all performance as it is brought to peak performance. The cost benefits analysis presents one operation as a snap-shot in time to present the cost of one ship-set of product.

The cost numbers converted and collected are combined in a single document incorporating the feasibility and cost benefits analysis reports. This document and the data it contains is passed to the leadership for assignment to a designated accounting person or team to investigate, validate, and certify the calculations for inclusion in the development of an ROI analysis and report.

5.2.2.6 Return on Investment

Once accounting has certified the calculations contained in the cost benefits analysis report, the ROI development can begin. The total cost of the machine, including ancillary cost, is combined into one number.

Hours to perform a one-ship-set hand operation for D/C are subtracted from the hours to perform a one-ship-set automated operation for D/C of the designated holes.

If this number is negative (no savings from automation), the process stops. There is not an ROI. The feasibility and cost benefits analysis reports are "shelved." An out-brief is provided to management to present the data and results.

However, management may decide to continue with the project for other reasons than economic incentive to reduce cost. Quality improvement and other positive benefits may entice and incentivize leadership to pursue the installation if the loss is minimal.

If the hours for the hand operation are greater than the hours needed to perform the automated operation for D/C of the designated holes, then there is a positive difference between the two operations.

The difference in hours is converted to dollars or euros (or other currency as applicable). Hand hours per ship-set (HHPS) – Automated hours per ship-set (AHPS) = Ship-set savings in hours (SSH). For example, if it takes 100 hours to D/C the designated holes by hand and 50 hours to D/C the same holes by automation, then: (100–50) = 50 hours of savings per ship-set.

SSH are multiplied by the savings dollars or euros. If this amount is \$55 or €55, then the dollar savings per ship-set would be (50 × 55) = 2750 dollars or euro per ship-set.

This amount is now multiplied by the airframe quantity produced per year. For example, if there are 50 airframes produced per year, then the annualized savings would be (2750 × 50) = 137,500 dollars or euro per year.

This annualized number is divided into the total of all costs to bring the automation online. If the machine for this application were to cost 1,000,000 dollars/euros, then the ROI would have a (1,000,000/137,500) = 7.273-year payback. Most companies require a three- to five-year payback, so in this example, the ROI would not warrant a recommendation for implementation.

In most cases, the ROI does warrant a go ahead based on a positive economic outcome. In some cases it does not.

If it does not, then it does not mean the demise of the potential project. Occasionally, an assessment to expand the envelope of the machine to capture additional holes or make the machine portable to other assemblies or products can enhance the ROI. An example of a portable machine that was designed to facilitate increase mobility is the MOGADS machine described and illustrated in Chapter 4.

5.2.2.7 Cost Benefits Analysis Deliverables

The deliverables of the cost benefits analysis include the feasibility report, cost data, cost benefits analysis report, ROI, and recommendation for go/no-go of the project.

The purpose of the deliverables is to convey the necessary information to leadership to make an informed and logical decision based on objective data.

5.3 "Go-Ahead" Decision Criteria

The results of the study are compiled and combined with the cost benefits analysis results in a report and presented to production leadership. Production and executive leadership determine the fate of the project. While the data contained in the cost benefits analysis report play a crucial role in the company's leadership consideration and decision to automate the D/C operation, other considerations may come into play. Other considerations can positively affect a negative ROI and drive a decision to go ahead with the project; and other considerations can negatively affect a positive ROI and result in a decision to cancel the project.

Many of these "other considerations" are contained in Chapter 3.

5.4 Summary

Once the decision has been made by a company's leadership to consider automation as a viable alternative to D/C an airframe by hand, a detailed process is set in motion to facilitate the gathering of objective data.

The objective data gathered from the process is compiled and summarized in a document that provides company leadership with the opportunity and economic potential for replacing hand operations with automated operations.

The document includes the results from a feasibility study, cost benefits analysis report, and the ROI potential. It also identifies the cost centers where application of automation has the greatest chance of success should the decision be made to proceed.

If the decision is made to proceed with automation of the D/C process, Chapter 6, How to Choose the Right Automation, and Chapter 7, Steps to Installation, will help guide the reader through the necessary steps for success.

Chapter 6
How to Choose the Right Automation

There are no safe choices. Only other choices.

—Libba Bray

6.1 Introduction

The first step to choosing the right automation for D/C of airframe assemblies is a requirements document. There are several automation manufacturers who can design and build a machine that will perform the operations necessary for quality holes. The requirements document communicates your desired outcome as to when the machine is installed, where it will be used, its integration requirements, and the complexity level of the tasks it will perform.

There are many approaches to writing a requirements document. The success or failure to achieve a satisfactory automated D/C machine begins and is guided by the contents of the requirements document. Extremely detailed information limits the supplier to providing you with exactly what you wrote/defined in the document and lifts the responsibility for performance from their shoulders. I have heard suppliers tell the writers of extremely detailed requirements documents when the machine failed or did not perform to their expectation, "I gave you exactly what you asked for."

Strategically partnering with selected suppliers who can help write and understand the outcome you want to achieve is a valuable action that results in a satisfactory outcome. You want to drill selected holes with countersinks on an assembly through a defined stack of materials. Working with the supplier design team to achieve the best requirements document to achieve the optimal result is beneficial to you and your suppliers. The degree of specificity is a joint effort aligned with the requirements of both the builder and the user of the machine.

This chapter will help you develop a template for a requirements document. It will also provide examples of both specific and boiler plate Requirement Documents that have been developed and are used by industry.

Selecting the right automated D/C system also involves understanding the environment that will surround and support the machine. The complexity of the operations it performs and the supporting infrastructure need to be analyzed and assessed to the level of capabilities that exist within the factory. Therefore, the requirements document must be aligned with the capabilities of the factory to support the chosen machinery.

Another consideration for choosing the right piece of automated machinery is the Keep It Simple (KIS) method of selecting machinery. Buyers often lose focus when confronting the choice between what the machine can do vs. what we want the machine to do. Requirements creep has caused the demise of many good-intentioned automated machine applications. It is better to be successful with the simplest application that can be economically feasible than to grow the requirements and complexity of the machine to capture a few more dollars of savings and fail.

In addition to providing guidance to develop a robust requirements document, this chapter will provide instruction in requirements creep avoidance and focused operational simplicity for success. Application of these principles will facilitate the acquisition of the simplest machine to install for your application to ensure success. It will also enable through modular requirements the expansion of the machine to include other enhancements once the machine is operational and has gained a reputation for success.

6.2 Three Don'ts

When beginning the quest to choose the right automated D/C machine, it is important to remember the three don'ts. Don't buy what you don't need. Don't buy what you can't afford. Don't buy what you can't handle. Keeping in mind the three don'ts before, during, and after the formulation of the requirements document will help minimize requirement creep.

Requirements creep is a constant threat to the project. It is rooted in the fascination with the technology and gadgets that modern automation machines can offer as add-ons and options. Add-ons and options increase the gap between needs and wants.

6.2.1 Needs

The needs that are the basic requirement for a D/C machine are to drill or D/C holes. Wants are those add-ons and options that can always be justified to enhance the performance of the machine. They most often enhance the performance of the machine minimally. The inclusion of a wide variety of factory personnel is necessary for inclusion, buy-in, and support of the automation project. The team approach to defining project requirement promotes requirements creep. Every group member in the complex world of airframe assembly sees their portion of the project differently. For example, the quality engineer wants automated inspection systems that reduce the inspector's involvement in the process. Adding automation enhancements complicates the machine and its operation, adding cost. While automating the inspection process is valuable, it adds a level of complexity to the primary purpose of the machine: D/C of holes.

The focus on ***just*** drilling or D/C is essential to the success of the automation project for first-time buyers. It is even important for the buyers of a second machine, where the success of the first drove a desire for a second machine. If all the machine does is drill or D/C holes, the project will be a success. That is the purpose of the machine and that is what will define a 100% successful project. Not one person will stand back and say, "It doesn't drill holes very well. And it took longer to get up and running. But man! Look at that tool changer!"

As the machine matures in its operation, add-ons and options can be purchased and installed as improvements during regularly scheduled periodic maintenance.

Prevention of requirements creep takes strong team leadership. There are pressures from senior leadership, stakeholders, team members, and others who will ask the question, "Could the machine do this?" The answer from the team leader should be yes, but this machine will not.

Machine suppliers will introduce team members to the latest gadgets through direct contact. Engaging the suppliers and integrators and integrating them into the project team is the best approach to defining requirements. But their participation puts them in direct contact with individual team members who often are exposed to Cadillac enhancements that benefit the supplier more than the project. A firm stand from the project lead with support from their leadership promotes focus on the intent of the project, which is to drill or D/C holes. The internal and external customer's primary desire is for holes, acquired in the simplest and most cost-effective way.

Figure 6.1 shows a highly complex piece of automated D/C machinery with every enhancement available on the market at its time of purchase. Much of it is not used, and it complicated its acquisition, installation, and operation, slowing its transition to production.

Fig. 6.1 A highly complex piece of automated D/C machinery.

6.2.2 Affordability

Purchasing the best machine that money can buy is a great idea for Bill Gates. It is not a very good idea for companies trying to hold down cost. Defining precisely what the machine is to accomplish and working with a supplier/integrator to understand the entire life cycle cost is the first step to hold down cost. In their zeal to promote automation, internal and external proponents of on-assembly automated D/C machinery often focus on front-end costs, and they limit the focus specifically to the machine. When considering machinery for automating a process, the machine's total impact to the cost of surrounding operations, the machine's Life Cycle Cost (LCC), infrastructure changes, and facilities modifications to support its operation must be included.

Figure 6.2 shows the relationship model of the difference between the purchase cost of a machine and the total life cycle cost, installation, and integration cost. The picture graphically illustrates the hidden costs that are often overlooked when considering a new piece of machinery.

Buying a slightly less efficient machine (slower drill rate) could result in a smaller overall cost. In addition, its efficiency rate could improve through operating simplicity. Purchasing F-35s is analogous to purchasing an automated D/C machine. The cost of each airplane is approximately $100 million. The total life cycle cost is projected to be $800 million, an 8:1 ratio [6.1]. Complex machines are not any different than the F-35 example. Their purchase price can be the tip of the iceberg.

Moderate Analysis	Cost Center	Number Of Holes		Baseline	Alternate	Delta Savings	Machine Cost	ROI
	A	3000	0.8	24000	19200	4800		
	B	1500	0.8	12000	9600	2400		
	C	4500	0.8	36000	28800	7200		
	D	1800	0.8	14400	11520	2880		
Total		10800	0.8	86400	69120	17280	5,000,000	≈289 Units
Conservative Analys	Cost Center	Number Of Holes		Baseline	Alternate	Delta Savings	Machine Cost	ROI
	A	3000	0.7	24000	16800	7200		
	B	1500	0.7	12000	8400	3600		
	C	4500	0.7	36000	25200	10800		
	D	1800	0.7	14400	10080	4320		
Total		10800	0.7	86400	60480	25920	5,000,000	≈192 Units
Agressive Analysis	Cost Center	Number Of Holes		Baseline	Alternate	Delta Savings	Machine Cost	ROI
	A	3000	0.6	24000	14400	9600		
	B	1500	0.6	12000	7200	4800		
	C	4500	0.6	36000	21600	14400		
	D	1800	0.6	14400	8640	5760		
Total		10800	0.6	86400	51840	34560	5,000,000	≈145 Units

Fig. 6.2 Automated machine cost-based relationship model.

The cost of the machine can include inflationary cost that is directly attributable to the inexperience of the factory personnel tasked with supporting the acquisition team. Their fears and misconceptions can drive cost due to non-value added activity and acquisitions. These include adding complexity to the machine and its operational processes to assuage their fears. The training and orientations described in Chapter 10, Training, can minimize or eliminate the costs associated with automation phobia.

Another activity that can drive cost is "too much machine" for the job. Owning a machine with a Siemens 840D controller when a simpler controller will meet your needs adds expense and complexity, or buying a machine with extreme precision derived from expensive software and hardware can add 20 to 25% to the acquisition cost and 50% more to maintenance and calibration costs.

One consideration that can add considerable cost is the financial viability of the companies you are evaluating and partnering with to develop a requirements document. After winning the $18.9 billion contract to build the Joint Strike Fighter (JSF), Lockheed Martin Aeronautics Co. in the summer of 2002 placed a $12.3 million order with Ingersoll Milling Machine Co. for custom-made automated D/C machines to produce airframe assemblies. As of April 2002, Lockheed Martin had paid Ingersoll more than half the contract price [6.2]. A giant in the machine tool field, Ingersoll was one of five companies in the world capable of manufacturing the sophisticated automated D/C machine needed for the F-35. The Rockford, Illinois, factory had sent signals of possible trouble in the years before 2002, when it sold off several divisions. But the cessation of operations came as a surprise that sent a shock wave through the aerospace industries.

Lockheed Martin Aeronautics' concern about any delay in delivering three highly advanced custom drilling machines for its F-35 Joint Strike Fighter convinced the Illinois bankruptcy court to allow it to hire competitor Cincinnati Lamb (Hebron, Kentucky) to complete the machines while various suitors, including Camozzi, made bids for Ingersoll [6.3]. An assessment of the financial viability of the machine manufacturer would have saved the buyer Lockheed Martin millions of dollars and prevented project delays.

Assessing the financial health of a company can be a very complex procedure. Many elements contribute to the financial health of a company. Western accounting refers to the accounting methods and practices used in the western-most countries of the world as opposed to those located farther to the east. Generally Acceptable Accounting Principles (GAAP) and International Fair Reporting Standards (IFRS) codes have many similarities, but also many differences. For example, when assessing the financial health of a company in Spain, caution should be used. A company might be financially strong and have a positive financial position, but viewed from the perspective of the United States accounting practices, the evaluation outcome seems negative.

6.2.3 Complexity

Operating complexity also contributes to buying the right machine. Chapter 5, Considerations before Replacing Hand Drilling with Mechanization or Automation, describes very important considerations before selecting the right machine for your factory application. Machine precision, complexity, and operating architecture influence the size of the factory infrastructure needed to support the machine.

Arriving at a facility in Florida to assess the viability of installing automated D/C, we left after a month with the recommendation that we should not automate the airframe. This was not because there were not many great opportunities to apply simple, effective, and low-cost automated solutions. There were. The factory did not have any infrastructure in place to install, integrate, make operational, and sustain automation. The cost to bring the infrastructure up to the standard necessary to support automated D/C systems would be prohibitively high. The factory facilities and operations could not support automation.

They could, however, support low-cost mechanization of D/C that required minimum expense to modify their factory facilities and supporting infrastructure.

Two takeaways from the Florida experience. One was that not all opportunities are what they seem. On its face, the opportunity was apparent. Digging deeper revealed more complexity to the decision to automate. The other was that there may be alternatives to improve a D/C operation such as mechanization when automation is not feasible.

6.3 References

Before writing the requirements document, a good way to determine what works best is to visit airframe assembly operations where on-assembly automated D/C machines are used. There are opportunities for the asking.

The best place to engage others who have experience with automation is at conferences such as the Aerospace Manufacturing and Automated Fastening (AMAF) Conference held by SAE International, the Defense Manufacturing Conference (DMC), or the Aerospace Automation Consortium. They usually have tours associated with the conferences that provide further opportunities to network and engage those with experience in automation. Most of the professional aerospace societies such as the Society of Manufacturing Engineers (SME) and the Society of Automotive Engineers (SAE) have LinkedIn discussion groups to facilitate dialogue. Just like any other product, the best view to the success of a product and the people who supply it is through those who have already purchased and are using it.

6.4 Requirements Document

Once you are ready to tackle the hard work of developing a requirements document, there are decisions to make. There are three approaches to writing a requirements document. One is the minimalists view. Provide the supplier/integrator with the engineering and tooling models, explain what you want to accomplish, and ask for three proposals to accomplish the objective. This view takes into account that there is limited number of aerospace on-assembly D/C suppliers. The limited number of suppliers makes it fairly certain that they know the speeds and feeds requirements for the requirements of the airframe. Although there are a limited number of suppliers, their approaches to solving the challenges vary. The solutions that you receive as a user will provide a wide assortment of solutions with the open architecture requirements document. The minimalist approach to a requirements document opens the product engineering, manufacturing processes, and tooling philosophy to the supplier. Supplying the engineering and tooling models as well as manufacturing processes is required for the supplier/integrator to understand the challenges, limitations, and controls necessary to provide the optimum solution. Giving up engineering and tooling models and manufacturing processes causes concern that proprietary data may migrate to competitors. There are also export control laws and the International Traffic in Arms Regulations (ITAR) that are United States Government regulations that control the export and import of sensitive military data and information. They also control the export of civil aviation data and information that might be transferable to military applications or restricted export countries. There are export control laws that also govern European and Asian nations as well.

The opposite view is the maximalist view, where a user believes they are the best judge and expert with regards to their tooling, engineering, and manufacturing processes. The maximalist believes the requirements document should be as robust and detailed as possible, defining everything from listing laws, rules, and regulations for conformance to the size and type of bolts used in the machines construction. The maximalist keeps the engineering and tooling models and all the manufacturing process information closed. Only questions asked that are relevant to the construction of the machine are answered, and the decision as to the relevance of the question is made by the user. This view to the requirements document drives very limited creativity and is change averse. The approach sensitizes change, slows the process, and increases cost. It is a highly defined document that is used as a tool by the supplier to supply only what is contained in the requirements document, and nothing more without engineering change authority, documentation, and compensation.

The best requirements document incorporates the two into a flexible compromise. It is a compromise between the limits of the company to reveal enough information to facilitate innovative and responsive design, and the need to protect sensitive proprietary or governmental data and information.

At a minimum, the requirements document should contain a basic template of information from the potential buyer to the supplier/ integrator. It should include the project title and relevant contact points for administration and technical questions for both parties. It should also contain general information, proposed operation summary, functional requirements, performance requirements, additional systems requirements, equipment and software definition, and security.

The fidelity and detail under each category of the template is a decision the each team acquiring automation must make based on internal criteria and expertise.

6.4.1 General Information

The general information section of a basic requirements document should contain the purpose, scope, project references, acronyms and abbreviations, points of contact, and information concerning coordination and timing of the project.

6.4.2 System Summary

The system summary section generally describes the current or pending operations, processes, controls, tooling, rate, materials, and factory operating environment.

6.4.3 Functional Requirements

The functional requirements that the machine will perform such as D/C 5500 0.190-in-diameter holes through aluminum to aluminum stacked material, 0.250 in thick are described. Of course additional functional requirements for the machine would be included for the unique application the user intends.

6.4.4 User Impacts

As experienced machine designers, builders, and integrators, the development, operation, and organizational impacts to the users should be described. Included in this section would be items such as power requirements to the machine or communication protocol for the controller.

6.4.5 Performance Requirements

This section describes the manufacturing processes and specifications including accuracy, validation, timing, capacity, and safety.

6.4.6 Additional System Requirements

Additional system requirements that are not defined should be included to enable the factory to prepare for the machine. These include system description, integration, commercial and customized equipment, and flexibility.

6.4.7 Equipment and Software

Equipment, software, interfaces, and communication requirements that the machine will require must be identified. Machines are placed into the assembly line where power, electrical, computer lines, and other connections need to be made for the machine to function.

6.4.8 Security

Security issues involve both control of suppliers within the user's facility and intellectual property. One addresses the issues associated with supplier's employees working and interacting with the machine user's employees. The supplier has the responsibility to define to the user who will be on their company property to ensure compatibility, reduce labor issues, certify citizenship (if necessary), and provide proof of insurance.

6.4.8.1 Factory

The requirements document has to ask that the proposal submitted by the supplier/integrator will define how much the machine installation and integration will disrupt the existing operations of the factory. The installation and integration crew from the supplier will need space for their equipment, tools, inventory, and people. A time, space, support, facilities requirements request should also be included as part of the requirements document.

6.4.8.2 Intellectual Property

Intellectual property security issues arise from the supplier/integrator's intimate understanding and exposure to the user company's manufacturing technology and processes. It also affects the customer of the machine's user. The requirements document should contain security requirements such as nondisclosure agreements and security clearances (if needed) as part of the requirements document.

6.4.9 Location

The requirements document should also define the supplier's location and support during the machine installation and integration. Geographically diverse suppliers, some located offshore from the company using the machinery, have a duty and responsibility to provide localized personnel. This requirement should be spelled out and a staffing plan requested to support the local operations.

6.5 Summary

A well developed and appropriate approach to selecting an on-assembly automated D/C machine can result in a positive outcome. Figure 6.3 shows three complex on-assembly auto-D/C machines attached to an assembly tool. The machines drill both sides and the upper section of an airframe independently or concurrently. The system operates in conjunction with a highly active cell where humans and machine interact.

Fig. 6.3 Highly complex three on-assembly auto-D/C machines attached to an assembly tool.

A poorly written requirements document that is either too vague or too detailed can cause poor results. To begin, the machine user's team must focus on the ability of the factory to support a piece of automation, at the right price, for a specific purpose. Requirements creep is a disease that can derail a well-planned project. A strong project lead has to maintain the focus on the primary function of the machine, which is to drill or D/C holes.

The requirements document will be different for every application. However, at a minimum it must include the primary elements of general information, system summary, functional requirements, user impacts, performance requirements, as well as equipment, software, and security requirements.

Once the proposals are in, the team should look at the proposals to determine the best machine to perform the requirements defined in the document. They should also evaluate the supplier's capability to financially build and sustain the highly specialized machinery over its entire life cycle.

References

6.1 Pentagon F-35 Fighter Under Fire in Congress, PBS, http://www.pbs.org/newshour/bb/military/jan-june10/defense_04-21.html, Retrieved June 9, 2012.

6.2 Ross, Paul, "Machine tool co. bankruptcy sends ripple through JSF program," *Aviation Week*, July 17, 2003.

6.3 http://www.thefreelibrary.com/Ingersoll+MT's+aerospace+contracts.-a0113337655">Ingersoll MT's aerospace contracts, Retrieved June 11, 2012.

Chapter 7
Steps to Installation

The first rule of any technology used in a business is that automation applied to an efficient operation will magnify the efficiency. The second is that automation applied to an inefficient operation will magnify the inefficiency.

—Bill Gates

7.1 Introduction

There are many steps to successfully install any piece of modern numerically controlled machinery. The complexity of the operation performed by automated D/C machines integrated into a production line populated by centers designed for hand operations complicates the process steps. The steps necessary for successfully installing and integrating a complex piece of automated D/C machinery include the entire environment surrounding the factory and its operation. One reason this is essential is that the nature of on-assembly D/C involves drilling through a collection of parts of differing material types, varying thicknesses, produced by a variety of methods, at many locations. All of these contribute to a diverse collection of parts assembled by hand in jigs and fixtures routinely calibrated once a year.

The skin of the airplane is placed over the collection of parts and called the substructure. When D/C is performed, the driller, whether human or machine, cannot see the exit point of the drill when it leaves the substructure component. The human drilling process mitigates the guess work through a process called step drilling. Placement and trim variations are accommodated through adjustments based on exit measurements to ensure ED conformance. ED conformance is one of the critical specifications necessary for the structural integrity of the airframe. ED specification conformance is not a validation of part trim and placement conformance.

Automated D/C machines drill one-shot full-size holes at the design-specified vector location. The machine assumes the trim, part placement, and other variables are within the confines of the specified tolerance and therefore have an exit hole that meets ED specification.

The first step in the automation process is to evaluate the existing hand process, part placement, and trim process for conformance to specifications. Many bad holes have been drilled by automated machinery with the assumption that holes drilled by hand and having acceptable ED validates the substructure part placement and trim specification conformance. It does not. In addition, if validation is not performed prior to machine operation, then the machine is the first suspect under the rule of, "all was fine until we started drilling with that machine."

The hand method of D/C covers a variety of sins that are invisible until automation is applied. Therefore, the first step in the automated D/C machine installation and use process is to validate the trim and part placement conformance to specification.

This is just one example of the additional steps necessary to ease the application and ensure a smooth application of automated D/C into the airframe assembly process.

The process of evaluating and adjusting the process to ensure part placement and trim conformance will be described with other steps that are unique to the success of automated D/C airframe assemblies. The unique steps specific to automated D/C on assembly involves the entire production process. This includes tooling and the means of production as well as honest evaluation and corrective action to stabilize production processes previously assumed to be in control. This chapter will include the predicated steps necessary before automating the D/C operation in an assembly center.

7.2 Mapping the Existing Process

Before the requirements document is created, the existing process is mapped. Process mapping is performed within the cost center where automation will be installed. Process mapping is also performed upstream and downstream of the cost center where automation will be installed. Upstream and downstream process mapping is necessary because the application of on-assembly automated D/C alters the process flow into and out of the cost center. Quality, schedule, and cost effects are identified, measured, and adjusted to accommodate the installation of the automation. For example, the quality of parts entering the cost center must conform to new trim requirements to ensure proper edge distance. The assemblies leaving the assembly cell affect the cumulative assembly variance when aligned with other assemblies. The variance may affect the fit between a hand-produced (D/C) assembly and an automated assembly in which holes were drilled by machine.

The process map also includes the external supply chain. The effect of automation on the supply of parts and components from external sources has to be mapped, evaluated, and adjusted for.

Process mapping is also known as process charting or flow charting. It is one of the oldest, simplest, and most valuable techniques for streamlining work. It is also subtle and requires experienced facilitators for best results

A process map visually (through simulation and flowcharts) depicts the sequence of events to build a product or produce an outcome. For automation application it also includes additional information such as cycle time, inventory, and equipment information. In most automation situations the process map:

1. Follows one assembly and the assembly's component group.
2. Identifies the critical dominant component(s).
3. Shows other components and subassemblies.
4. Describes each event concisely.
5. Adds additional information as required.
6. Merges items on the process map when they are joined as an assembly.

In many cases a planning document exists, and using the existing document to incorporate into the process map is tempting. The process plan is a guide to the mechanics and in many cases does not represent the actual events occurring when parts are fabricated and collected into an assembly. The performance of a process map also offers the opportunity to evaluate and incorporate lean principles into the process [7.1].

During the process of mapping the current state, a broad-based team from several levels is assembled. It includes mechanics because they know the details of what really happens. It also includes engineers and support people because they have a broad view of the process as it is supposed to be performed.

Manufacture of complex structures involves various processes that yield the final product. Each process step from the beginning of the process affects the material properties or behavior of the assembled parts and thus the impact to the effectiveness of the automation. Mapping and assessing the quality and behavior of an assembly or collection of assemblies arising from multistage manufacturing processes from the beginning to the end of the process is essential. The process map of the current manufacturing flow and recommendations to stabilize the process for automation is the deliverable of the team. The installation of automation into a predominantly hand-built assembly line amplifies the areas where processes are "dynamic." The installation and integration of an island of automation into the hand-built process causes the bridges into and out of the assembly cell to become choke points if the dynamic hand process is not brought into control. Identification of areas where the variability (dynamics) of the hand process must be corrected is a part of the process map deliverable of the team. Three essential areas of focus for evaluation and potential adjustment are part placement, trim, and finished hole size.

7.2.1 Part Placement

A critical element of the process map is measurement, evaluation, and recommendation of corrective action necessary to tighten and hold parts in a repeatable position.

The detailed components of an assembly tool are made to locate and hold the airframe component parts of the assembly into their specified positions. The tool details that hold the parts in position are designed and built to specified weight. This is done to allow humans to lift, locate, and bolt them to the assembly tool. These tooling pieces are called removable details. Every time an assembly tool is readied for the next assembly, the removable details are assembled inside the tooling frame. This process requires the mechanic to precisely place and bolt dozens of steel removable details together that in turn are meant to hold the parts in position for drilling, countersinking, and fastening.

When the parts are assembled and fastened together, the removable details are disassembled to allow the assembly to be removed from the tool. The assembly and disassembly of removable tool details occurs each time a new assembly is produced. Figure 7.1 shows a typical assembly tool with the associated collection of removable details attached.

Fig. 7.1 Typical assembly tool with the associated collection of removable details attached.

Tools are made and annually calibrated by the tool fabrication group to meet a standard removable-tool-detail specified tolerance of ±0.010 in (±0.254 mm). While the tools are manufactured and calibrated to the specified tolerance, the daily shop operating environment and other conditions contribute to an actual tolerance variance greater than the ideal. The specified true hole position of each hole in the assembly is typically ±0.010 in (±0.254 mm). The assumption in the process, made when holes exit the substructure and meet the edge distance tolerance, is that parts have been placed within the specified tolerance. This illusion is perpetrated by the mechanic's ability to "pull" the hole into conformance through progressive steps in drill size. The only way to ensure the proper position of the tool details and parts is through the prohibitively expensive process of checking (laser, photo, or optical measurement) of the tooling details at each assembly.

Two ways to correct the variations in the tooling removable-detail assembly process is replicating the hand process of shifting the holes; or collectivizing the various removable details into on-piece tooling for part detail location.

Automating the hole shifting process requires machine software and a vision system for the machine. In this process the substructure is assembled and clamped into the assembly tooling prior to applying the skin. The substructure is mapped by the vision system, and the software optimizes the hole drilling program to reposition any hole that, when drilled, would not meet the edge distance criteria. This process is being used in a number of factory applications of automation. The idea of the substructure varying from the specified engineering allowable tolerance is unacceptable to some engineering and production operations leadership.

An alternative to shifting holes at every assembly is to collectivize all the removable details into a single, solid, one-piece detail to hold all the parts. The collectivization of details creates a heavy tool that now replaces the human assembly of the tool details with a single large assembly requiring a crane or mechanized medium of transport. The collectivization eliminates the tolerance variance associated with the assembly and disassembly of dozens of pieces of tooling.

Figure 7.2 shows a single-piece mechanized tooling detail containing all the details formerly independently assembled by hand.

Fig. 7.2 Single-piece mechanized tooling detail containing all the details formerly independently assembled by hand.

7.2.2 Trim

Parts are trimmed with consideration to the design specification. The distance from the bottom of the part, where it is typically placed against the tool locating detail of the assembly tool, and the top of the part, where it is trimmed, affects the ED. The tolerance for trim can vary from ±0.040 in (±1 mm) down to ±0.010 in (±0.254 mm). A part bought off as acceptable to trim tolerance, when added to the part placement tooling detail tolerance, exceeds the acceptable edge distance tolerance for hole placement. Figure 7.3 illustrates how the combined part tool tolerance and trim tolerance exceed ED tolerance.

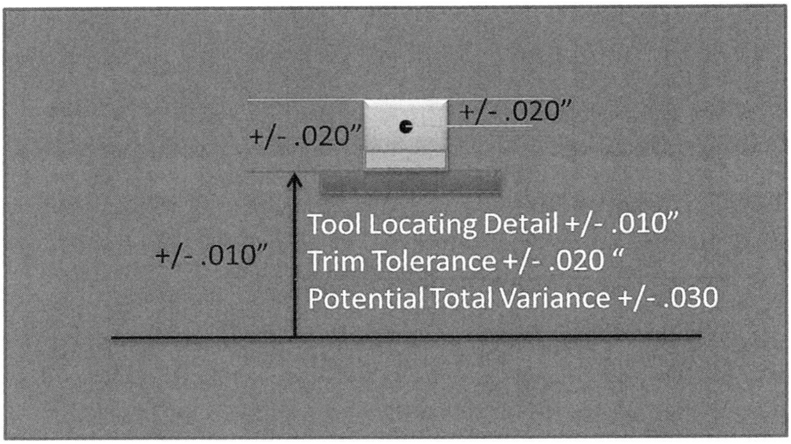

Fig. 7.3 Combined part tool tolerance and trim tolerance exceed ED tolerance.

Again, the mechanic's ability to shift the hole location visualizes the illusory effect that the hole is placed to engineering specification.

The rise of on-assembly automated D/C holes has led to trade studies to determine better trim technology to reduce trim variance. One method is to incorporate precision trim technology such as NC machines into the trim process. Other methods are to enclose the machines into environmentally controlled rooms (controls CTE), improve trim fixtures, and use machine precision enhancement software such as Volumetric Error Compensation (VEC). These technology enhancements for trimming parts provide essential reduction of trim tolerance.

7.2.3 Finished Hole Size

Drill bits for drilling holes are purchased within a tolerance band that allows for a quantity of re-sharpenings to maximize the ROI for the drill. The drill bit in the hand operation is inserted into a drill motor that is mostly pneumatic and hand operated. As a result, the hole sizes meet tolerance, but within the acceptable limits of variability. The hole size is usually at the upper control limit of the tolerance.

Fasteners are purchased at a specified tolerance and vary within the tolerance. Therefore, most fasteners will fit in a hand-drilled hole. Many fasteners will not fit in a machine-drilled hole. Machine-drilled holes are consistently the size of the drill. They are at the lower end of the control limit. Driving fasteners into a hole smaller than its diameter causes stress on the material surrounding the hole and in the case of composites material, has caused delamination.

Several alternatives have been used by automation integrators to solve this problem.

The acceptable tolerance on the fastener can be reduced to align it with the consistency of the automated drilled holes, or the drill bit size can be increased closer to the upper control limit of the hole.

One solution aligns the fastener size with the consistent hole size. The other increases the hole size to facilitate fastener variance.

7.3 Mapping the New Process

Once the old process is mapped, a new process incorporating the changes to the old hand process and incorporating additions to facilitate the new process has to be mapped. Figures 7.4 and 7.5 show the process flowchart for a typical military vertical stabilizer incorporating auto D/C with hand activities using unitized tooling moved by mechanized means.

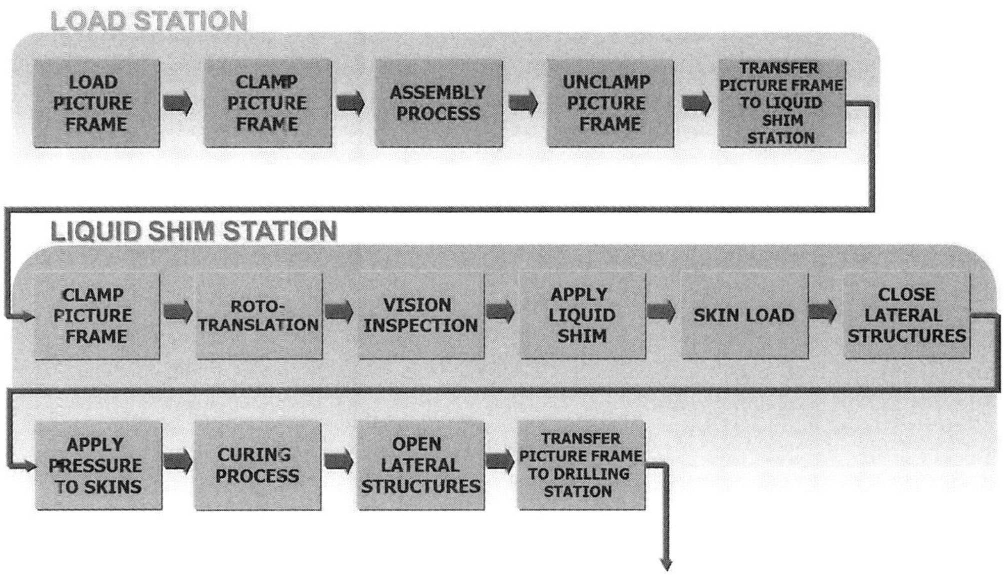

Fig. 7.4 Initial setup and operations in preparation for auto-D/C.

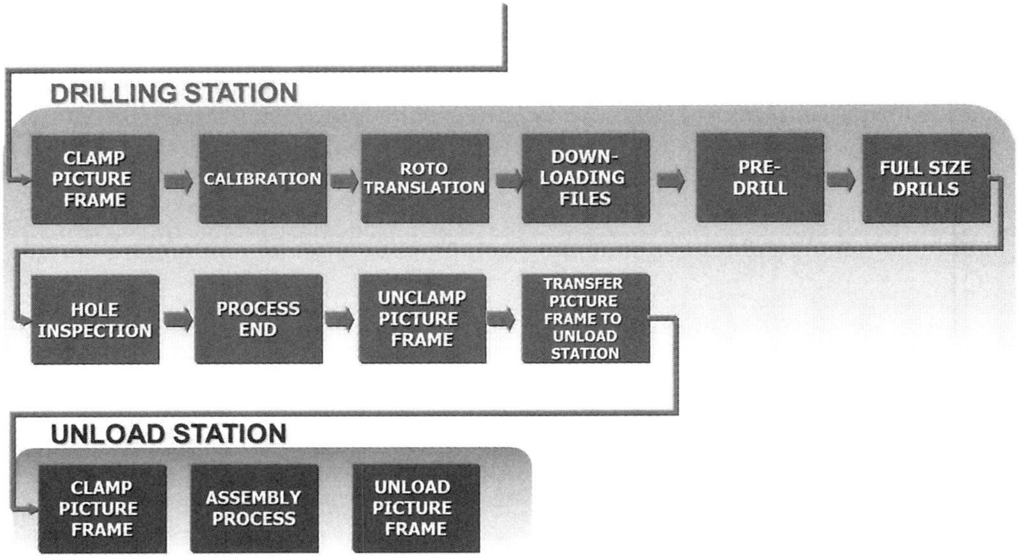

Fig. 7.5 Process flowchart for a typical military vertical stabilizer incorporating auto D/C.

7.4 Requirements Document

The key to any successful on-assembly automated D/C requirements analysis is to identify what the new system will do for all appropriate users and stakeholders, and what those users and stakeholders need it to do. Various techniques can be employed to ascertain requirements, which need to be clear, concise, and relevant to the organization. When employed properly, requirements that may conflict with each other can be discovered and resolved early in the process. Once the analysis is complete, a written document should be created that is the roadmap for addressing all of your organization's automation needs.

7.4.1 Identify Stakeholders and Define Stakeholder Needs

As a first priority, the project's executive leadership sponsor must be identified. It is essential that all stakeholders know who has responsibility for the project's scope [7.2].

Then identify the key people within the organization who will have a stake in the project, and the direct users of the platform. The project must meet their needs, and their inputs must be taken into consideration. Each and every stakeholder must understand and be able to articulate what they expect the automation platform to do and define their roles and responsibilities within the organization to understand how the various stakeholders will affect one another on a daily basis.

A responsibility assignment matrix (RAM) [7.3], also known as RACI matrix [7.4] describes the participation by various roles in completing tasks or deliverables for the project [7.5]. It is especially useful in clarifying roles and responsibilities in the cross-functional/departmental auto drill project and processes [7.6]. RACI is an acronym derived from the four key responsibilities most typically used: Responsible, Accountable, Consulted, and Informed [7.7].

Stakeholders are categorized by responsible, accountable, consulted, and informed.

7.4.1.1 Responsible

Individuals who are responsible to perform work to achieve specific tasks necessary for the success of the project are identified. There is typically one role with a participation type of *responsible*, although others can be delegated to assist in the work required.

7.4.1.2 Accountable

The accountable individual is ultimately answerable for the correct and thorough completion of the deliverable or task. The accountable individual delegates responsibility for its completion. An accountable person must sign off (approve) work the responsible person provides. There **must** be only one accountable individual specified for each task or deliverable.

7.4.1.3 Consulted

Consulted individuals are those whose opinions are sought, typically subject matter experts, and with whom there is two-way communication. A subject-matter expert (SME) or domain expert is a person who is an expert in a particular area or topic.

7.4.1.4 Informed

Informed individuals are stakeholders who are kept up-to-date on progress, often only on completion of the task or deliverable, and with whom there is just one-way communication.

Very often the role that is *accountable* for a task or deliverable may also be *responsible* for completing it (indicated on the matrix by the task or deliverable having a role *accountable* for it, but no role *responsible* for its completion (i.e., it is implied). Outside of this exception, it is generally recommended that each role in the project or process for each task receive, at most, just one of the participation types. Where more than one participation type is shown, this generally implies that participation has not yet been fully resolved, which can impede the value of this technique in clarifying the participation of each role on each task.

7.4.2 Categorize System Requirements

To facilitate analysis, requirements are categorized:

- Functional Requirements define how the system will function from the end-user's perspective.
- Operational Requirements define operations executed for the product duration and the ultimate workflows resulting from the project.

- Technical Requirements define the technical issues considered for successful execution.
- Transitional Requirements are needed to implement the new product or process smoothly.

7.4.3 Interpret and Record Requirements

Determine achievable requirements, and how the system can deliver them:

- Define requirements precisely to avoid ambiguity and transparency, and to ensure that business objectives are met.
- Identify the most critical requirements, and which are expendable.
- Fully understand the consequences the project will have for existing processes, systems, and staff.
- Resolve any conflicting requirements issues with stakeholders.
- Determine the reliability and usability of the system.

Results are compiled in a detailed report and distributed to leadership and major stakeholders. Feedback is solicited and a deadline assigned for final document compilation and signature.

7.4.4 Sign Off

Once the relevant feedback has been incorporated into the report, sign off from leadership of the stakeholders signifies their agreement that the requirements as documented reflect their input and expected outcome from the system performance. The requirements document sign off keeps the execution focused and reduces requirements creep.

7.5 Platform Specification

The platform specification incorporates the requirements document as baseline for the system definition. A successful approach is to use simulation to provide a notional approach to begin development of a more detailed platform specification. The notional simulation provides visualization of the operations of the machine that can lead to revision. They are less costly than simulation models, and they have more fidelity and can be used for an initial evaluation. They can also be used as a visual tool to promote understanding of the project and its operation. Figure 7.6 is a screen shot of a notional design for a generic vertical stabilizer drill/countersinking machine.

There is argument within the automation user community as to how detailed an automated platform specification should be. Too much information and the specification is too rigid, forcing the supplier to conform in areas where they may need flexibility. Too little information and the contract terms, definition, and technical specifications are left to the interpretations of the supplier.

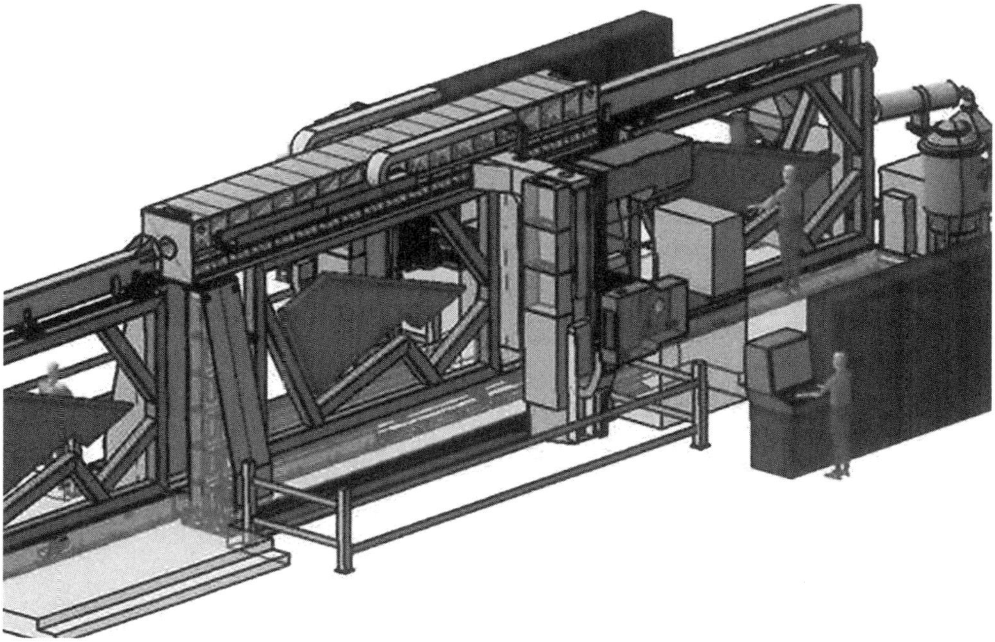

Fig. 7.6 Screen shot of a notional design for a generic vertical stabilizer drill/countersinking machine.

At a minimum, it should contain the requirement that the machine incorporate those attributes and technology necessary to conform to local, state, and federal laws and guidelines, including those that will pass into law during the life of the machine. While it is impossible to know what may become law, it is possible to know when a law will take effect and whether the law has an effect on existing technology or its operation.

Also, it should include the basic operating expectations of the machine and its integration with the facilities available for its operation. The platform specification for a piece of automated machinery is unique in that each is different. That is why suppliers/integrators should be included as stakeholders early in the process, preferably as part of the requirements document development team.

7.6 Request for Proposal

The Request for Proposal (RFP) incorporates all the former documents including the requirements document, process map (current and proposed), and the notional simulation. The documents are attached as supporting documentation to facilitate a clear understanding to the supplier of the intent and operational expectations of the auto-drill system.

The RFP also requests a detailed timeline for milestones, beginning at Purchase Order (PO) release:

- Engineering Design Review schedule at kickoff, 25%, 40%, 70%, and 100% final design buyoff.
    - As part of the Design Review, a detailed schedule for completion of engineering definition for assembly tool, load/unload station, transfer stations or mechanisms, and drilling station.
    - Electrical and Software Design.
- Material Acquisition, including long-lead-item release approval, long-lead items acquisition (spindle and tool changer), and materials acquisition.
- Factory Assembly (supplier) of the machine and loading station.
- Factory Acceptance Test (supplier).
- Transport and final assembly (user facility).
    - Packing and shipping.
    - Import considerations (where applicable).
    - Foundation (if required) need date.
    - Assembly.
    - Final acceptance and buyoff.
- Project management organizational structure for support of the project from the supplier team is also requested as part of the RFP.

The RFP defines the elements and format of the response to ensure uniformity from the respondents and ease of evaluation. At a minimum, the RFP should contain the following items.

7.6.1 Introduction

The introduction contains an executive summary off the supplier's response to the RFP. It defines the main concept developed to meet the requirements of the RFP and describes the proposed conceptual solution based on machines previously developed by the supplier for similar applications. The response should also include a set of assumption such as what information will be required to add fidelity to the cost and schedule if the supplier is awarded the contract. These include knowledge of the manufacturing/assembly processes, tolerances, type of material, thickness to drill, and disposition of the drills.

In addition, the introduction contains information on the background, objective, concept proposed, and system description.

7.6.1.1 Background

The background provides assumptions pertaining to the availability of information for which the proposal is based, such as assembly and delivery, tooling/drilling hardware, and engineering. The background describes the machine performance requirements to meet the requirements definition. It also describes the respondent's expertise and experience for installing similar machines to fulfill similar requirements. The background identifies the primary design system such as CATIA V5, flooring and foundation modifications if needed, caveats to the RFP, and recommended changes to the original RFP.

7.6.1.2 Objective

This section defines the objective of the supplier and presents their concept to demonstrate their proposal's alignment with the objectives of the RFP.

7.6.1.3 System Description

The system description demonstrates the understanding the supplier has of the interaction of their proposed solution with the RFP specifications, and the interaction with other operations in the work cell. The system description includes the mechanical characteristics composed of the drill/countersink spindle head assembly, axis of motion description, spindle type and specifications, coolant system, vacuum system, and tool monitoring system. The system description also includes the electric / electronic equipment composed of connection to the main power supply, electric cabinets, main operator panel, and electrical installation.

As part of the system description, the Computer Numerical Control (CNC) hardware includes software features; and characteristics of the numerical control, motor drives, and vision system are defined.

7.6.1.4 Cost Estimation

Cost estimations are provided with options in United States dollars (USD), euros, or other currency as requested in the RFP. It is in this section that the supplier designates their currency designation for payment.

7.6.1.5 Acquisition

Supplier selection is governed by different criteria within the internal governance of each company. Commercial and especially military procurement of machine tools and robots are also subject to rules, regulations, and laws of local, regional, and the country where the machine will be used.

The early integration of your supplier teams through strategic alliances facilitates the success of the project. Strategic alliances can also reduce the supplier selection frustration experienced when running the gauntlet of the complex local, regional, and country's rules, regulations, and laws governing acquisitions.

Strategic alliances are often accompanied by teaming agreements that spell out the terms of the alliance and smooth the transition to the procurement phase. The early Engineering and Manufacturing Alliance Project Teaming Agreement defines the terms of the agreement and can facilitate a sole source/directed source document to aid with specific supplier selection. The sole purpose of the early Engineering and Manufacturing Alliance Project Teaming Agreement is to serve the needs of the parties in their effort to work cooperatively on the development of an automated system.

Some of the issues affecting the procurement of machines across boarders are the exchange rates and their effects on the ROI of a machine tool. A machine purchased in the United States for delivery to Europe two years from now can have a profound effect on the cost of the machine upon delivery. A 2,000,000 USD machine purchased on contract today (May 28, 2012) dollars would cost 1,600,000 EUR at an exchange rate of 1$ for .8€. Delivered two years from now, if the euro continues to slide against the dollar, the billing rate for the machine at 1$ for .9 EUR would be 1,800,000 EUR, a €200,000 increase. Buying futures to hedge against currency fluctuations mitigates the effects. Insisting on a contract paid with the currency where the machine is delivered is another option.

7.7 Manufacturing Readiness Level Consideration

At this point in the process, the purchase order has been released and the detailed design is underway [7.8]. Material specifications are approved and materials are available to meet the planned pilot line build schedule. Manufacturing processes and procedures have been demonstrated in a production representative environment (simulation or lab), and detailed producibility trade studies and risk assessments are in process. Cost models have been updated with detailed designs, rolled up to system level, and tracked against targets. Unit cost reduction efforts are underway. The supply chain and supplier quality assurance has been assessed. Long-lead procurement plans are in place. Production tooling and test equipment design and development have been initiated.

A Critical Design Review (CDR) is now performed. The CDR is a milestone within the system development process. The on-assembly auto-drill system design is evaluated against requirements to verify the outcomes of previous activities and identify issues before moving forward with the project. The CDR identifies the progress of the project and identifies, clarifies, and reprioritizes tasks within the project to realign the schedule with expectations. The successful CDR will trigger the system release.

The auto-drill CDR is compulsory as part of the system design control when developing unique, custom-built equipment for use on highly regulated products such as aerospace airframe assemblies. By definition, a CDR must include persons who are external to the design team.

Most formalized systems engineering processes recognize that the cost of correcting a fault increases as it progresses through the development and installation process. Additional effort spent in the early stages of development and installation process to discover and correct errors is worthwhile. Design reviews are an example of such an effort. Therefore, a number of design reviews may be carried out, for example to evaluate the design against different sets of criteria or during various stages of the design process.

Other more detailed and continuous tracking tools ensure visibility and alert leadership and automation teams of areas critical for success that have drifted outside acceptable metrics.

7.8 Performance Measurement Tools

As the system develops in maturity and the elements for success become clear and defined, the application of performance measurement tools becomes important to ensure the success of the project [7.9]. One performance measurement tool is the Earned Value Management System (EVMS). It is a single integrated system able to provide accurate forecasts of project performance problems. EVMS has the ability to combine measurements of scope, schedule, and cost.

Recent research studies have shown that the principles of EVMS are positive predictors of project success [7.10]. New Federal Acquisition Regulations (FAR) and Defense Federal Acquisition Regulations (DFAR) include EVMS in part because EVMS can help substantiate contract disputes [7.11]. Popularity of EVMS has grown significantly in recent years beyond government contracting.

Essential features of any EVMS implementation include a project plan that identifies work to be accomplished; a valuation of planned work called Planned Value (PV); and predefined metrics to quantify the accomplishment of work, called Earned Value (EV). EVMS implementations for large and complex projects such as on-assembly automated D/C machine installation include many more features, such as indicators and forecasts of cost performance and schedule performance. However, the most basic requirement of an EVMS system is that it quantifies progress using PV and EV.

Using the EVMS system to manage the installation process of an on-assembly automated D/C machine can smooth the process and installation.

In the United States, the primary standard for full-featured EVMS is the ANSI/EIA-748B standard. The standard defines 32 criteria for full-featured EVMS compliance. Other countries have established similar standards.

7.9 Factory Acceptance

Once the machine has been constructed and is ready for acceptance, a Factory Acceptance Test (FAT) is performed at the supplier. The FAT is a customized testing procedure for the specific automated D/C machine that will be disassembled after test, packaged, and shipped to the factory where it will be used. The FAT is the final test of the machine performed after system debugging and capability testing, and before relative accuracy classifications are performed. In many cases it is difficult to predict the correct operation of the system or consequences due to failure. FAT is a valuable check before shipment. The FAT test cases are selected during the planning phase and incorporated as part of the test plan in the purchasing contract document.

The main objective of the FAT is to test the fully functioning instrumented system. The tests are performed during the final part of the design and engineering phase before the transition to production phase at the place of the system's use.

Planning procedures to carry out the FAT should be well defined and documented. A clearly defined set of tests should be formulated for the unique automated system.

The FAT test document includes the appropriate tests to be performed as well as the people who will have the responsibility, accountability, and authority (RAA) for developing the test cases. Because the testing activities are general methods, the FAT is applicable to programmable and nonprogrammable operation of the automated system. The most important part off the FAT addresses the specifications for test cases. Test cases must have clear definition and description, have well-structured procedures, and be relevant [7.12].

The FAT includes a test protocol defining who is responsible to validate and attest to the test outcomes. The RAA for each test case should not only have the appropriate level of competence and experience to evaluate the success of the test case, but also be independent of the acquisition team. They must perform an independent assessment and audit of the system to match system performance to the requirements document and the performance specification.

The minimum tests run should incorporate the basic process control system (BPCS), fault, failure, error, hazard, hazardous situation, hazardous event, harm, instruments, process risk, safety instrument function (SIF), and defined accuracy validation.

7.9.1 Basic Process Control System

This test determines the ability of the machine to meet the performance criteria set out in the requirements document as the objective of the machine. The determination of functionality is determined by test of the response of the system to input signals from the NC program, other associated equipment, other programmable systems (tool changer, pressure foot, etc.), and operator input. It tests the output signals causing the integrated machine process and its associated equipment to operate in the specified manner. This part of the test determines the basic operability of the components integrated as a single system. This part of the test is meant to discover fault, failure, or error.

7.9.1.1 Fault

Fault is an abnormal condition that may cause a reduction in or loss of the capability of the system to perform a required function.

7.9.1.2 Failure

Failure is the termination of the ability of the system to perform a required function.

7.9.1.3 Error

Error is the discrepancy between a computed, observed, or measured value or condition and the true, specified, or theoretical correct value or condition.

7.9.2 Hazard

The safe operation of the system is of primary concern and is an important part of the FAT. Hazard tests are meant to discover the potential sources of harm to the operator or others who may intrude into the operational region of the machine. This part of the test is performed after the successful BPCS. As part of the hazard evaluation and test, the machine is run through a routine performance program and observed to discover/identify a hazardous situation or hazardous event that could cause harm to a person. This test is meant to assess operational functions that will introduce a hazardous situation. Safety system tests are performed later in the FAT.

7.9.2.1 Hazardous Situation

A hazardous situation results in a circumstance where a person is exposed to hazards.

7.9.2.2 Hazardous Event

A hazardous event is defined as an event that would cause harm to a person.

7.9.2.3 Harm

Harm is the physical injury or damage to the health of people, either directly or indirectly, as a result of machine malfunction, safety system malfunction, machine operation, or damage to the environment or property.

7.9.3 Instruments

The instrument test examines the apparatus used in performing an action. This test validates (or discovers fault in) the functionality of the instrumented systems meant to interact with the operations of the automated machine. Tested devices include sensors such as pressure, flow, temperature, and position sensors; logic solvers and control systems such as programmable controllers and distributed numeric control (DNC) systems; and safety instrumented systems.

7.9.4 Process Risk

This test evaluates the potential harm to the product arising from conditions caused by abnormal events. For example, a mechanic has left a clamp in the way of the machine where the machine's programmed motion will impact the clamp. In this case, the test would assess the machine's impact sensors to determine its ability to recognize, stop, and notify.

7.9.5 Safety Instrument Function

The SIF tests the designed safety system's independent and integrated function to the safety integrity level necessary to achieve functional safety of the machine during operation and at rest. This test evaluates both the safety instrumented protection function and the safety instrumented control function.

7.10 Installation and Test

Installation and test at the point of delivery and use is the final step in the installation process. At this point the Transition to Production team takes over. Their RAA and actions are described in Chapter 9, Transition to Production. The machine installation is performed at the factory in cooperation with the acquisition team. The test and acceptance processes and procedures before transition to production are described in Chapter 8, Test Procedures.

7.11 Summary

The steps necessary up to and including installation are meant to provide a logical progression of activities to ensure that the system delivered meets the expected outcome.

These include mapping the existing process to determine areas that need to be addressed and corrected prior to installation of automated machinery. A "to be" map is developed from the existing process map to define the new automated D/C machine's interaction with the hand tasks performed in and around the automation. A notional concept simulation provides visibility and clarifies the new process intent.

A requirements document development team is identified, and those responsible, accountable, and with authority for actions for procurement and management of the system acquisition are assigned. Individuals who are stakeholders are identified and classified according to their participatory and reporting level. The sponsoring leadership is identified, project budget approved, performance measurement systems put in place, and the supplier integration begins.

The requirements document and technical specification are collected, and an RFP is issued to qualified suppliers.

Suppliers respond to the RFP, and supplier selection takes place with the issuance of a purchase order. Performance measures and milestones are formalized and integrated into the performance measurement tools to ensure project visibility. The completion of the final milestone for the machine build is the FAT that validates the performance of the machine and its operating systems at the supplier's factory.

After successful FAT, the machine is broken down, packaged, and shipped to the buyer's facility. At the buyer's facility, the machine is uncrated and inspected visually for damage. The machine is assembled at point of use and presented to the Transition to Production team to complete the operational integration into the assembly line.

References

7.1 Ruffa, Stephen A., *Going Lean: How the Best Companies Apply Lean Manufacturing Principles to Shatter Uncertainty, Drive Innovation, and Maximize Profits*, AMACOM. ISBN 0-8144-1057-X, 2008.

7.2 Allinio, Requirements Analysis & Workflow Design, http://allinio.com/services/automate/requirements-analysis-workflow-design/ Acquired 02/23/2012.

7.3 *A Guide to the Project Management Body of Knowledge (PMBOK Guide)*, PMI Standards Committee, Project Management Institute, ISBN 1-933890-66-5, 2010.

7.4 Jacka, Mike and Paulette Keller, *Business Process Mapping: Improving Customer Satisfaction*, John Wiley and Sons, ISBN 0-470-44458-4, 2009, p. 257.

7.5 Margaria, Tiziana, 4th International Symposium on Leveraging Applications, Isola Heraklion, Crete, Greece, October 18–21, 2010, Proceedings, Part 1, Springer, ISBN 3-642-16557-5, 2010, p. 492.

7.6 Brennan, Kevin, *A Guide to the Business Analysis Body of Knowledge (BABOK Guide)*, International Institute of Business Analysis, ISBN 0-9811292-1-8, 2009, p. 29.

7.7 Blokdijk, Gerard, *The Service Level Agreement SLA Guide—SLA Book, Templates for Service Level Management and Service Level Agreement Forms. Fast and Easy Way to Write Your SLA*, Lulu.com, ISBN 1-921523-62-X, 2008, p. 81.

7.8 "Manufacturing Readiness Level (MRL) Deskbook," Version 2.0, May 2011. Prepared by the OSD Manufacturing Technology Program In collaboration with The Joint Service/Industry MRL Working Group.

7.9 Marshall, Robert, "The Contribution of Earned Value Management to Project Success of Contracted Efforts," *Journal of Contract Management*, 2007, pp. 21–331.

7.10 "New federal contracting rules can help or hurt says former contracting officer," *Huntsville Times*, 23 September, 2011, http://www.al.com/42/index.ssf/2011/09/new_federal_contracting_rules.html, Retrieved November 15, 2011.

7.11 "U.S. Penalizes Huntington Ingalls," Reuters, November 11, 2011, http://www.aviationweek.com/article.aspx?id=/article-xml/awx_11_11_2011_p0-393081, Retrieved November 15, 2011.

7.12 *Factory Acceptance Testing* Guideline, Process Industry document IEC 61511. 25, August 2006.

Chapter 8
Test Procedures

Building technical systems involves a lot of hard work and specialized knowledge: languages and protocols, coding and debugging, testing and refactoring.

—Jesse James Garret

8.1 Introduction

Automated D/C machines placed into use on the assembly line cannot use the standard circle and cube acceptance tests for positional accuracy and repeatability validation performed on off-the-shelf machinery. The machines dedicated to D/C thousands of holes precisely and normal to the surface they are drilling through need additional criteria to determine their acceptability prior to and after installation. Holes are spaced of varying sizes, usually less than an inch apart. Each hole is an opportunity to create a defect, and, therefore, the testing regime includes the individual components of accuracy at the hole, normalcy to the surface where the hole will be drilled, and compression deflection resulting from compound contributors. The repeatability of the process is also a critical part of the test and validation acceptance test. The deviation from the first run and a set of data over several runs are examined to determine if drift occurs, and if it does occur, if it is within predetermined acceptable criteria.

Standard machines are built in factories where the environment is similar to the environment in which they will be used; therefore, factory acceptance tests are a good medium to validate the performance and accuracy of the machine.

Automated D/C machines are built in the same environments as standard off-the-shelf machines, but are applied into assembly lines where a very different environment challenges the machine's performance by the exclusion of cleanliness, sound foundations, and vibration.

The same FAT test is performed on automated D/C machines as a preliminary step to the tests that are performed after the machine is installed at the user's facility.

The tests performed are more detailed and include tests under load and stress to determine their performance during operation, and not just their conformance to positional accuracy.

This chapter will provide the steps and the unique tests performed on automated D/C machines for quality assurance and confidence prior to D/C on the airframe. The tests described will include safety testing. Testing of safety devices is essential for the operation of machinery in an environment where humans work in and around machinery. Safety testing also includes those operations that detect and shut down machines that malfunction and could cause harm to the product. Testing and validation of safety systems for personnel and product protects the two most valuable assets of the company.

In addition, environmental testing processes will be described to determine the Volatile Organic Compounds (VOCs) emitted by the machines that are not relevant or measured in the hand operation. Machines use lubricants for D/C that are in mist form and emanate into the environment. The rapid D/C of composite components and skins combines with the coolant mist to enter the atmosphere as VOC contaminants. Extraction systems, which are attached to the machines, remove the contaminants to levels below the defined limits. Therefore, the acceptance tests must include testing criteria within the factory where they are used to meet the specific state guidelines.

8.2 Unique Tests

There are unique tests that are run to validate an on-assembly D/C machine and that are not performed on any other machine type. They are run to validate the various actions and functions of the machine in an effort to replicate as close as possible the first article run of the machine. The tests are run at different levels of fidelity to progressively combine the actions of the machine. They conclude with several full runs to confirm that all complete actions of the machine are validated. These are the last tests to be run before the machine produces holes and countersinks in flight hardware.

Each of the tests may use a separate NC program to minimize the time of the test. Efficiencies are derived from only using the elements of the fully operational NC program needed for the test.

8.2.1 Impact Avoidance Test

The maxim, "The shortest distance between two points is a straight line" is the most dangerous approach to NC programming for an automated D/C machine. There is a lot of "stuff" between point A and point B.

Motion to, between, and from hole position is fraught with danger. The coordination between the NC programmer, tool designer, and manufacturing engineer is essential to reduce the risk of collision while optimizing efficiency. The tooling and parts challenge the NC programmer to leave the home position, transit to the hole position, perform the required activity, and move to the next hole. This is all done while avoiding the part and tool. Simulation software such as DELMIA for manufacturing can help. The software tools virtually define, plan, create, monitor, and control the machine motion to optimize performance and reduce the risk of errors such as collisions. Machine path simulation has become a major contributor toward reducing NC programming errors and minimizing collisions. The merged models that contain the tooling and parts engineering combine with the machine architecture to facilitate a virtual view for planning motion. Once completed, it is a valuable tool for team collaboration to assess and evaluate the process sequence. Figure 8.1 is a picture of simulation path planning for collision avoidance.

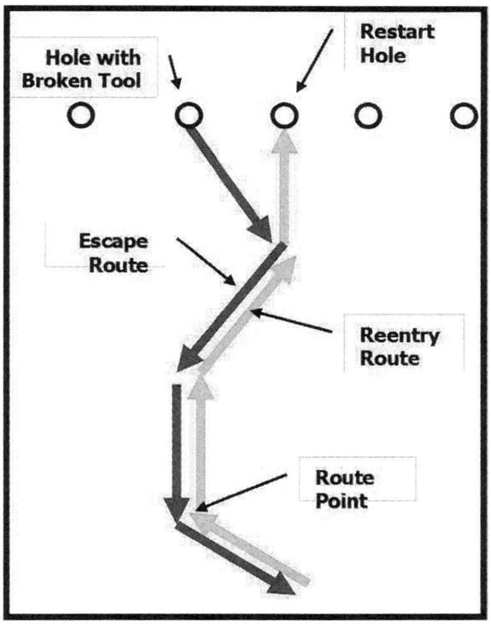

Fig. 8.1 Picture of simulation path planning for collision avoidance.

Automated D/C machines perform numerous operations. An airframe assembly may have as many as forty different-size holes, some with and some without countersinks, and some cold worked. The holes are usually drilled through stacks of different materials of varying thickness requiring different drills, speeds, and feeds. The optimal sequence to minimize motion and maximize output is facilitated by a combined team to perform simulation analysis. The sequencing includes the standard tool engineers who must plan for the tool (drills and countersinks) quantities, tool holder types and quantity, their order in the tool changer, and tool set lengths. Sequencing also includes the determination for placement and planning for temporary fastener installation. Temporary fasteners help hold the skin and structure together during the one-shot D/C operation.

The virtual world is a great beginning but does not always represent the as-built world, where on-assembly automated D/C is performed. As-built assembly tools often differ dimensionally from the designs. The large structure is welded together and heat treated. Rough structure transitions to precision in the tooling-details mating surface, where parts are placed. In between rough structure and mating surfaces lie many deviations and design changes. The changes are not always captured in the simulation model. The changes may not even make it to the design model. Paper changes by a liaison engineer to facilitate a request by the manufacturing engineer or tool builder may wait for batch processing into the computer model. Batch processing is done when convenient. Handwritten design changes also lose some of their fidelity because of the handwriting, level of detail, and time.

It is important to run the machine to assess the degree of disconnects between the virtual world and the physical world. Over reliance on simulation and collision avoidance systems can lead to major negative events. Running impact avoidance tests with the machine is the only way to evaluate the simulation model and NC program and avoid the risk of collision derived from assumption. The collision assessment test is also a way to train operators and gain confidence in the performance and safety of a machine.

Collision avoidance tests are performed with the tool fully loaded. All the parts of the assembly and the tooling are installed when the test is run. The collision avoidance test is run very slowly.

Collision avoidance tests should be run *eyes on*. Eyes on means getting a team of people to observe every part of the machine that moves. They watch for clearance or potential obstructions. It is especially important to get up close to the head of the machine. At a minimum, the person watching the head of the machine should have a remote controller that can stop or slow the machine as needed. The remote controller is in addition to the observation team having radio communication devices, as the machines are large and factories are noisy. The placement of an operator/observer at the controller assures backup and access to further machine emergency controls and stops. The controller also has images provided by the camera that is on the head of the machine to enhance observation.

As part of the collision avoidance test, the collision avoidance technology, operating in conjunction with the machine, is tested. Within the dynamic environment of airframe assembly, people operating in proximity to machines are a safety concern. People operating in the proximity of automated D/C machines also inadvertently attach or leave items such as tools, clamps, padding, air lines, drill motors, fans, sealant guns, and other objects that can interfere with or damage a machine. A pre-operation walk-around and checklist to facilitate the thoroughness of the walk-around helps to remove these items. However, the size of many on-assembly automated machines can defeat the best observation techniques.

Running collision avoidance tests causes anxious moments, because the best way to confirm the collision avoidance ability of the system is to place an obstruction in the machine path and see if it stops before impact. The anxiety is reduced or eliminated by use of soft, light objects that are not fastened down. They offer enough pressure or size for the system to recognize the obstruction without actually harming the machine. I have used rubber drill motors and foam clamps placed where it would be anticipated that a mechanic would overlook their removal. Figure 8.2 shows a collision avoidance system that would be mounted on a machine.

Fig. 8.2 Collision avoidance system.

It should be noted that as part of the collision avoidance assessment test, the drill penetration point and extension of the drill point through the substructure should be evaluated. When doing the ink dot test, looking where the drill exits the material provides insight into possible tooling and hardware impacts that might occur. Many a hole has been ruined and drill broken during the first run because everything had been checked except this important item.

8.2.2 Aluminum Tape Ink Dot Test

I do not know of any other machine that uses an aluminum tape ink dot test to validate the alignment between machine, parts, and tooling; as well as validate the sequence and performance of the NC program. The test also validates the alignment of the angle of the drill head centerline with the surface of the skin. A unique NC program expedites the test regime and results in a visual record for evaluation. As the name implies, a ball point pen cartridge is chucked up into a tool holder. Aluminum tape is placed on the structure where the pen tip will strike the structure or skin. There have been many attempts to replicate the pen cartridge as a tool for this type of inspection, but all are expensive and none are better than a cheap BIC cartridge. Figure 8.3 shows a simple pen cartridge used in the aluminum tape ink dot validation test.

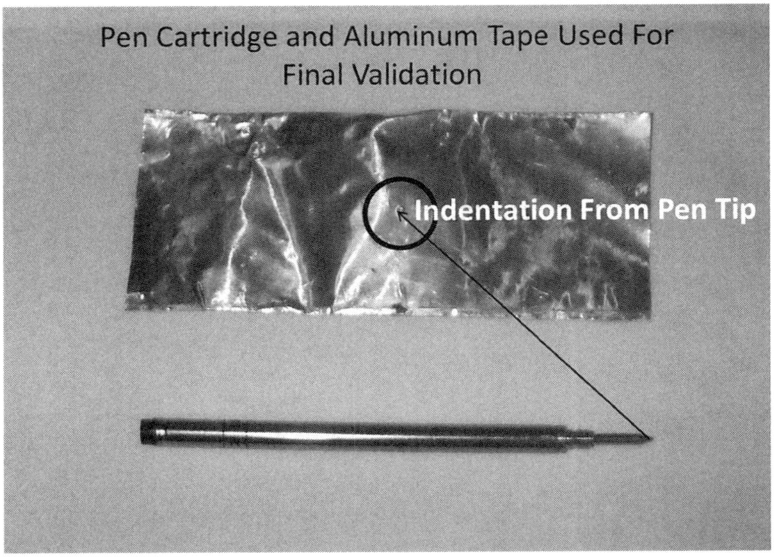

Fig. 8.3 Pen cartridge used for ink dot validation test.

Automated D/C machines use a unique system to locate the skin of an airframe. The need to seek and find the skin of an airframe prior to D/C is necessary due to an accumulation of variance caused by the assembly of substructure parts, skin thickness variance, materials used between the skin and substructure, and tooling variance. The location of the skin is transferred to the controller, which then activates the D/C routine. The drill/countersink tip is preset to a precise distance from the face of the pressure foot. The drill/countersink routine can be activated, resulting in speeds and feeds for each type of material in the stack at that location. And the proper depth of countersink will result.

The design of the end effector for D/C uses an electromechanical system to tie the face of the end effector to an encoder. A preset compression window is set in the encoder. On the GADS, the compression window is set at 25 counts. The encoder registers 1000 counts per inch or 1000 counts per 25.4 mm. When the machine arrives at a hole position, the X, Y, Z and rotational axis are locked. The pressure foot face is usually ~0.5 in or 12.7 mm off of the skin surface. This distance allows enough safety clearance for assembly variation. Once the X, Y, Z and rotational axis is locked, the command is given to the quill to begin motion to approach the skin slowly. The rate of approach is 25–50 in (635–1270 mm) per minute. When the pressure foot face touches the skin, it begins to compress the encoder. When the encoder reaches the assigned window, a signal is sent to the controller to begin the subroutine for the specific operation, including speeds, feeds, and depth control to begin. In the case of the pen dot test, the subroutine would spin up the rpm and feed rate, extend the quill until the pen momentarily touched the tape, and then rapidly retract. Once fully retracted, the X, Y, Z and rotary axis would be unlocked for movement to the next position. Figure 8.4 shows the encoder pressure foot relationship.

Fig. 8.4 Encoder pressure foot relationship.

8.2.2.1 Skin Off

The skin off ink dot test is performed to evaluate the relationship between the position of the hole the machine will drill and the substructure. The test is run on fly-away structure that is temporarily clamped into position. The test is normally run through a series of subassembled units in which the subassembly is disassembled and reassembled. The assembly and disassembly of the parts helps to validate the repeatability of the assembly process in which positioning the parts is critical to good ED. It also identifies the areas of trim where the trim may cause marginal ED.

Evaluation of the dots by the inspection team is done using the same method as when checking the ED of holes. The team measures from the center of the hole to the edge of the structure. In the ink dot test, they measure from the center of the dot.

The ink dot test is also used to look at potential design issues/errors where holes could exit into a radius.

Figure 8.5 shows an airframe with aluminum tape covering areas where the ink pen will leave a mark to confirm drill position on the airframe. In this view, the test is ready to perform with the skin on.

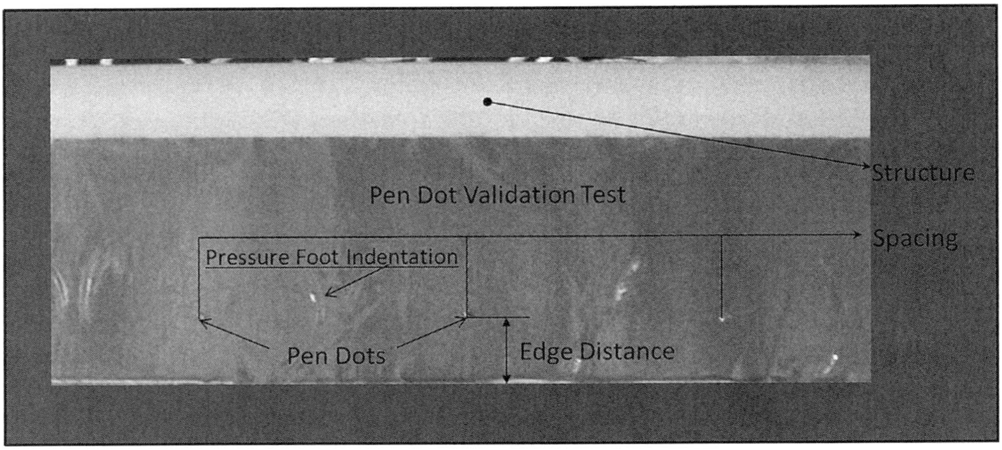

Fig. 8.5 Airframe with aluminum tape covering areas where the ink pen will leave a mark.

All of the information obtained from the ink dot test is used to modify tooling, if needed, to improve substructure location or repeatability, improve trim, and to revise engineering.

8.2.2.2 Skin-On

Skin-on tests are run the same as skin-off tests, using the same NC program. The purpose of the skin-on test is to view the indentation of the pressure foot into the aluminum tape to determine if the commanded position of the rotary axis is normal to the surface of the skin. This is critical to attain a flush countersink and prevent gapped fasteners.

The pressure foots radius is larger than the countersink head by a ratio of ~10 to 1. The larger radius amplifies error in the angle of incidence. Therefore, hole angle errors can be identified, analyzed, and corrected.

8.2.3 Compression Deflection Test

During the skin ink dot touch off test, the machine's pressure foot is performing a second function. It is providing pressure against the skin for several reasons. They include compression of the skin against the substructure and vibration reduction during D/C.

Compression of the skin against the substructure is necessary to prevent separation of the skin from the substructure when the drill breaks through the skin and enters the substructure. Separation occurs when the skin is released from the drill pressure and the pressure is transferred to the substructure. The separation, even if slight, can cause burrs and debris to build up between the skin and substructure. The burrs, and especially the debris, can damage the underside of the skin. It can also damage liquid shim and other materials that reside between the skin and substructure.

When the substructure is composed of harder material types such as titanium, the chip spiral produced by drilling can rotate between the skin and substructure and cause damage.

The pressure foot presses against the skin, providing compaction between skin and substructure to eliminate the challenge.

The pressure foot and drill pressure of the machine result in higher loads against the skin and substructure than a human could exert. Tooling for human drilled and countersunk holes are constructed with supports and clamp points dispersed to reduce deflection of the part and tool resulting from lower pressures. When automation is applied, the higher pressure results in deflection that requires rethinking the positioning of tool details. Too much deflection can result in elongated holes. Measuring deflection at the point of machine pressure can reveal where added details should be placed.

8.2.4 Drill Routine Conformance Test

Every time the machine stops and proceeds to the hole, a specific set of precisely controlled actions must take place. Confirming that a machine's speeds and feeds are correctly aligned with the position and specification for the hole to be drilled at that position is critical. Pecking actions, drill depth, hole size, and other actions are hole-specification dependent. The tedious task of sustaining a hole-by-hole physical observation and log is essential to prevent damage to a part. A polycrystalline diamond 0.250 diameter drill bit driving through composite skin and hitting titanium substructure at 4500 rpm and an 18 in/min feed rate is a recipe for a bad hole [8.1]. Worse yet is the sound it makes when it breaks in the hole and alerts the factory of the resultant defect.

NC programming, simulation, and modeling can and do make mistakes. The operation that a D/C machine performs is repetitive. Therefore, the added expense of spending time validating each hole's specific actions is warranted to prevent a defect the first time the machine rolls into action.

8.2.5 Countersink Depth Test

If drilling a hole using automated machinery is hard, then countersinking a hole with automated machinery is hardest. Drilling a 0.190–0.250 in (4.826–6.35 mm) diameter hole in aluminum or composite material, if properly lubricated and with optimized speeds and feeds, is like a knife moving through butter. The sharp, angled tip enters easy, cuts the material, and exits easy. Countersinking in the same material when the countersinking operation follows and is included as part of the operation increases the level of difficulty. Contributing to countersink process difficulties include size, angle, and debris.

A 0.190-in (4.826-mm) hole requires a 100° countersink with a diameter of 0.360 in (9.144 mm). The feeds and speeds must be changed and the pressure foot has to maintain pressure, after the drill breaks through the backside of the material stack, to dampen vibration when the larger countersink cutter of greater angle begins cutting. When the countersink cutter reaches the bottom of its cut, it has to dwell momentarily to clean out the hole. The pressure from the cutter pushing against the structure stops when the feed stops. The pressure foot must sense for the backpressure to not allow oversizing of the countersink. All of this makes for a complex operation that is compounded by residue on the surface of the skin remaining from the previous countersink. The machine's ability to control depth is derived from the face of the pressure foot and its relationship to the set length of the cutting tool, in this case the countersink.

Having debris on the surface to be countersunk defeats the operation of the machine unless added technology can remove the debris. The pressure foot uses a system from supplied shop air to blast the debris away before it contacts the surface.

All of these complex simultaneous and dependent activities make it imperative that the system be adjusted and tuned before full-size countersinking is performed.

The only way to validate the countersink system and "settle it down" is to countersink undersize holes with the machine during the first few units. The countersink depth is gradually increased to optimize the system and validate its capability. Until the fifth or sixth unit, the mechanics finish the countersinks and bring them to full depth by hand. While this may seem inefficient, the bulk of labor is in the initial removal of material in the countersinking process. The hole is cleaned out afterward by the mechanic. It is also less efficient than bringing the machine online and dealing with oversize countersinks.

The gradual progression toward full-size countersinks is frowned upon by product delivery team members, who want to get the machine into full operation as fast as possible. Resistance to these pressures will lead to their greater satisfaction later.

8.2.6 Machine Subroutines

There are subroutines on machines that increase the efficiency of the machines. All of these need to be tested to validate their operation and integration into the machine's function. Automated D/C machines have increasingly incorporated a tool changer. In the early days of trying to incorporate automated D/C onto the assembly line, suppliers, researchers, and designers would try to change the entire end effector each time they wanted to change the drill size. The end effector cost at the time was $250,000 US. So a drill/countersink setup with 16 different size holes would cost $4,000,000 US in end effectors. The move toward changing the drill in the complex end effector reached the application stage when, in 1997, the first automated D/C machine was installed on a military airframe. It was much less expensive to change the tool holders that ran $150 US. Kits were set up before each run of the machine, where 16 holes cost $2,400 US. The next generation of the machine incorporated a tool changer to further automate the system and take the human out of the loop.

The next-generation tool changer also incorporated a tool-set length validation system that provided a second check to ensure that the set length was correct.

All of these advancements should be tested on station to ensure their function and that the tool change path maximizes efficiency while minimizing impact potential.

Figure 8.6 shows a tool changer with set-length validation software.

Fig. 8.6 Tool changer with set-length validation software.

8.2.7 Final Full Function

Once all the components have been tested independently, they are put together and tested as a fully functioning system. All of the elements of the system are performed as dry runs to ensure access to structure, drill/countersink routines, tool change requirements, collision avoidance systems, and the operator work as a unified system. Dry runs are performed many times to gain confidence in the system. Confidence is built through repetition.

8.2.8 Demonstration

Part of the final flight hardware validation tests are demonstrations. When the tests are performed, the actions of the machine are observed as well as the outcomes. Many of the tests such as the skin-on ink dot and fully functional tests are performed for validation as well as demonstrations. The machine's fully operational demonstration to leadership and other stakeholders validates their support, promotes its capabilities, and instills confidence in the performance of the technology.

Demonstrations can be disruptive to the transition to production and product delivery teams. Due to the importance of the individuals that are invited to witness the automated machine "go through its paces," scheduling takes coordination. And the automated machine will remain idle before the demonstration event while awaiting their arrival. Internal and external members of the press who arrive with an entourage of photographers are also a part of demonstrations.

As disruptive as they are to the continuity of the testing schedule, they are an important part of the final validation of the machine. Demonstrations are visual representations of investment and risk that have come to fruition. Leadership and stakeholders promote the new automated D/C success by pointing to the technology as an example of the company's (and their) efforts to improve quality and reduce cost to the customer.

8.3 Volatile Organic Compounds

Environmental testing within the factory for VOC contaminants are a concern within the factory with D/C with automation. The end effector contains a vacuum to collect dust and mist overspray and deposit them in a container for later disposal. Figure 8.7 shows a modern centrifuge high-volume dust collection system.

Fig. 8.7 Centrifuge high-volume dust collection system.

In the United States, different regulations exist per state. Therefore, the state or country that the machine operates within has regulations for limits and testing methods defined for VOC testing.

Emissions testing is typically the most preferred source of emissions data from a specific source. Test data is a good representation of emissions from the source if the test is performed correctly and the emissions are reported in the correct units [8.2].

8.4 Summary

All of the tests described are the last tests before transitioning to production of an on-assembly airframe D/C machine. The tests are part of the final steps to validate the intricacies of the relationship between machine and human processes. These include the tooling, machine, and parts that have to be harmonized and aligned to achieve success. Other tests have to be performed early in the manufacture of the machine and at its completion; these include the FAT at the supplier and upon receipt at the user's facility. These tests are fairly common among machine tools and include integration and system's validation tests as well as tests to confirm the accuracy requirements of the machine.

No other machines need the type of testing that is required by the unique application and tasks required by an on-assembly automated D/C machine.

They include touch-off of the substructure and skin using testing methods that show visible evidence of the machine's activity. These tests use a simple method incorporating the advanced technology of the end effector combined with a simple ball point pen cartridge combined with aluminum tape to enhance the visual acuity of the inspectors.

All of the tests are run to assess and provide feedback for enhancement of the overall integrated system of tools, parts, and machine.

Due to the complex nature of the hand and machine operating relationship, compounded by the interaction and dynamics of their tasks, collision avoidance is a critical consideration for validation. While simulation and modeling help with collision avoidance and process sequencing, true validation can only happen in the physical world. The machine's collision avoidance system's ability to detect and avoid a collision in practice is the final adjustment needed after the virtual world has validated the machine motion in theory. The practical test in the physical world adjusts *should be* characteristics of operation to the *as is*.

Final validation of the machine includes the touch of skin and structure, compression, deflection, drill routine conformance, countersink depth, end effector subroutines, final full function tests, and demonstrations.

It surprises some that these tests far outpace the resource requirements traditionally required to test and validate an off-the-shelf or even many custom machines. It is the unique nature of the on-assembly automated D/C machine, and its integration into the airframe assembly process, that make these tests necessary. Some of the most costly components are combined in a collective of parts that result in very large investments. An on-assembly automated D/C machine can scrap very quickly if not properly tested before it touches fly-away parts.

Successfully tested, the machine becomes a testimony to the drive of the company to improve quality and reduce cost. The demonstrations that are interspersed throughout the final tests offer leadership and stakeholders to bear witness to the success of the project. They are often disruptive. The demonstration events need to be scheduled to minimize the impact to testing. They are worth the effort. The confidence and support that demonstrations foster in the leadership and stakeholders provide recognition, confidence, and expansion of the technology.

References

8.1 Garrick, Richard, *Drilling Advanced Aircraft Structures with PCD (PolyCrystalline Diamond) Drills*, Precorp Inc., Utah U.S.A., ATC-18, SAE International, 2007.

8.2 Pjetraj, Michael, *Influence of VOC Measurement and Reporting Methods on Regulatory Policy and Emissions Estimations*, NC Division of Air Quality Stationary Source Compliance Branch, December 6, 1996.

Chapter 9
Transition to Production

A skilled Transition Team leader will set the general goals for a Transition, and then confer on the other team leaders working with him the power to implement those goals.

—Richard V. Allen

9.1 Introduction

Transitioning one or more pieces of automated D/C machinery to the production floor is the most difficult part of the process for installing and using automation to D/C holes. All through the justification, selection, acquisition, test, training, and installation phases, "real" product was not involved. Airframe assemblies are comprised of expensive parts. Collectively, they represent thousands, and in some cases millions, of dollars of collective cost. A single hole improperly placed or a machine malfunction can destroy months of work, scrap a $100,000 assembly, and stop a production line. Often, the first hole that automated machinery drills and countersinks is in an airframe assembly. There are no test pieces that have been used to validate the process and that can be scrapped as part of the production readiness process. Therefore, it is imperative to the success of the machine's first operation and the health of the product that a robust technology and manufacturing readiness plan, risk mitigation process, and developmental roadmap precede the first series of holes drilled and countersunk on the airframe.

This chapter will identify the steps necessary as part of a production readiness plan to reduce the risk and some of the anxiety associated with pushing the button and D/C of the first hole in an expensive airframe. Nothing can eliminate the anxiety of watching the first, and each successive, hole being drilled as the machine moves incrementally along the airframe from position to position for the first time. If the steps described in this chapter are followed, anxiety will be reduced and most of the risk mitigated.

Included are recommendations for team member identification, risk reduction methodologies, risk reduction matrices, and test and validation techniques. The test and validation technology and techniques described include passive and benign methods that can be used on the actual airframe without risk to the product. These passive and benign techniques and technologies ensure and provide confidence to the TTP team, leadership, operators, and the customer that every aspect for machine performance has been satisfactorily tested prior to the drill touching the product.

A detailed and comprehensive matrix composed of a checklist of steps necessary to accomplish progressive transition to production is necessary. The checklist populates a risk reduction chart that identifies specific risk reduction efforts assigned before operational application of the automated D/C machinery can be brought online. The risk reduction charts and risk cube are used to provide visibility for presentation and communication to leadership and the customer of the actions of the team as they collectively progress towards implementation. Visibility and communication during the progression through the steps to implementation assuage the concerns of leadership and the customer.

Identifying a TTP team as a separate autonomous team and operation from the disciplines who have participated for requirements definition, acquisition, FAT, calibration, programming, facilities prep, and installation is essential to success. The TTP team provides the final production readiness step for signoff and validation prior to the drill touching the product.

9.2 TTP Document and Transfer of RAA

Transition to production culminates in acceptance of RAA by the Product Delivery Team for the operation and maintenance of the new automated D/C machine. The RAA is transferred from the TTP team to the Product Delivery Team. The Product Delivery Team includes stakeholders in the successful operation of the machine to perform the tasks outlined in the TTP document. The document contains the requirements and detailed steps necessary to reach the Technology Readiness Level (TRL) and MRL for its operation as an integrated part of the cost center. The document contains a roadmap that identifies critical milestones, timeline, and risk mitigation plan for their completion. The TTP document is signed by all core members of the TTP team and the Product Delivery Team when all milestones have been completed to bring the machine online. Due to the finality and significance of the TTP document, a Delegation of Authority (DOA) signature is not allowed on the TTP document for any of the Product Delivery Team. The TTP document is signed in a meeting (as opposed to a circulation signing) with the core RAA Production Delivery and TTP team leadership present. Once signed, the TTP document is replicated and distributed to each member of the Product Delivery Team, the TTP team, and their leadership. The original signed copy is incorporated into the build book for the automated drill/countersink machine and archived in a locked secure file within the Manufacturing Engineering or Manufacturing Technology department.

The signing (sign off) of the TTP document is the final step for factory acceptance of the automated D/C machine by Product Delivery. Therefore, the document including details, critical milestones, and timeline to be developed, compiled, agreed to, and signed is the first step for Transition to Production.

Before the TTP document can be initiated, the Transition to Production and Product Delivery Teams must be selected, orientated, and trained. The formation and activities of the core transition to Production team begins at the end of Technology Readiness Level and Manufacturing Readiness Level 4 and at the beginning of TRL/MRL Level 5.

9.3 TTP Technology Readiness Levels

Technology Readiness Levels are a systematic metric/measurement system that supports assessments of the maturity of a particular technology and the consistent comparison of maturity between different types of technology. The general model includes (a) "basic" research in new technologies and concepts, (b) focused technology development addressing specific technologies for the identified applications, (c) technology development and demonstration for application before the beginning of full system development, (d) system development, and (e) system "launch" and operations.

9.3.1 TRL 4

TRL4 includes the component and/or breadboard validation in the ***laboratory*** environment. It follows successful "proof-of-concept" work, where basic technological elements have been integrated to establish that the "pieces" will work together to achieve concept-enabling levels of performance for a component and/or breadboard. The validation supports the concept that was formulated, and it is consistent with the requirements of potential system applications. The validation is relatively "low-fidelity" compared to the eventual system: it is composed of ad hoc discrete components in a laboratory. For example, a TRL 4 demonstration of a new "fuzzy logic" approach to NC programming for automation might consist of testing the algorithms in a partially computer-based, partially bench-top component demonstration in a controls lab using simulated machine inputs.

9.3.2 TRL 5

TRL 5 includes component and/or breadboard validation in a ***relevant*** environment. At TRL 5, the fidelity of the component and/or breadboard being tested is increased significantly. The basic technological elements are integrated with reasonably realistic supporting elements. The total applications (component-level, subsystem level, or system-level) are tested in a "simulated" or somewhat realistic environment. From one to several new technologies are involved in the demonstration. For example, a new type of laser guidance system promising higher accuracy, validated at this level, would be used in an actual fabricated machine capability demonstration. The laser positioning system would be integrated with elements such as power supplies and supporting structure, and would be tested in a shop environment with simulation capability. At the close of TRL 5, the TTP and Product Delivery Teams have been fully integrated for the automated D/C machine factory operation.

9.3.3 Combined TRL/MRL

Industry has been using TRLs to measure technology maturity for many years. Recently, the use of TRLs has also been embraced by the U.S. DoD, as documented in the current version of the 5000 series publications.

While TRLs have proven useful in evaluating a technology's performance, as demonstrated in the laboratory or in a test environment, TRLs do not measure whether the technology product can actually be produced in an affordable manner. The concept of manufacturing readiness levels has been invented to expand the TRL idea so that it can incorporate producibility concerns.

9.4 TTP Manufacturing Readiness Level

The MRL approach addresses producibility concerns early in the development phase. MRLs are a metric used to assess the system engineering/design process and maturity of a technology's associated manufacturing processes to enable rapid, affordable transition to acquisition programs [9.1].

MRLs are quantitative measures used to assess the maturity of a given technology, component, or system from a manufacturing perspective. They are used to provide decision makers at all levels with a common understanding of the relative maturity and attendant risks associated with manufacturing technologies, products, and processes being considered. MRLs define the current level of manufacturing maturity, identify maturity shortfalls and associated costs and risks, and provide the basis for manufacturing maturation and risk management.

The Government Accountability Office has described it as best practice for improving acquisition outcomes [9.2]. In 2011, consideration of manufacturing readiness and related processes of potential contractors and subcontractors was made mandatory as part of the source selection process in major acquisition programs [9.3, 9.4].

9.4.1 MRL 4 Material Solutions Analysis, Milestone A

MRL 4 [9.5] decides the capability to produce the technology in a laboratory environment. Required investments, such as manufacturing technology development, are identified. Processes to ensure manufacturability, producibility, and quality are established and determined to be sufficient to produce technology demonstrators. Manufacturing risks are identified for prototype build. Manufacturing cost drivers are identified. Producibility assessments of design concepts are completed. Key design performance parameters are identified. Special needs are determined for tooling, facilities, material handling, and skills.

Examples of applicable action criteria for an automated D/C machine system at MRL 4 would be the successful completion of:

- Define true position accuracy classifications and assign risk
- Define system requirements and configurations for production drilling cells
- White paper solicitations for supplier/integrator approach to achieve various positional accuracy classifications
- Evaluation of technology development maturity and path forward
- Perform reach and line-of-sight analysis relative to production drilling cells
- Metrology analysis for guidance, coordinate reference systems, and inspection
- Define accuracy improvement options and perform cost analysis
  - Precision hardware
  - Software compensation
  - Metrology guidance
- Define preferred system concepts
- Finalize approach and refine list of precision components
- Define product oriented work breakdown structure (WBS)
- Define functional responsibilities and potential subcontractors
- Finalize proposal and capture funding

9.4.2 MRL 5 Technology Development, Milestone B

MRL 5 defines the capability to produce prototype components in a production environment. The required investments have been identified. Processes to ensure manufacturability and quality are in place and are sufficient to produce technology demonstrators. Manufacturing risks have been identified for building the prototype. Producibility assessments of design concepts have been completed. Key design performance parameters are identified. Special needs are identified for tooling, facilities, material handling, and skills.

Examples of applicable action criteria for an automated D/C machine system at MRL 5 would be the successful completion of:

- Precision hardware
  - Requirements definition and design relative to robotic platform
  - Fabrication of precision components as necessary
  - Order/receive COTS components, including robot platform
  - Install and test precision components
- Volumetric Compensation Software (VCS)
  - Define VCS process flow and requirements
  - Define software controls definition and enhancements
  - Development and test of VCS enhancements
  - Create VCS compensation and validation procedures
- Metrology components
  - System evaluation relative to platform capability and requirements
  - Metrology system process flow and requirements
  - Define software controls definition and enhancements
- Test of metrology guidance and validation procedures

The TRL and MRL system of evaluating each phase of progression is important to provide a clear path to success and maintain a communication medium for each step of maturity.

Figure 9.1 shows a representative TRL/MRL template for automated D/C transition to production.

9.5 TTP and Product Delivery Team Composition and Selection

The TTP and Product Delivery Teams are composed of two groups: a core group of experts that provides the leadership for completing the transition to production of the machine and a support group of individuals that provides expertise and guidance to the core group as needed. The core group has the RAA for the successful completion of the machine transfer to production. The RAA includes all project management including technical and budget. The core group identifies key contacts to populate the support group and obtains the necessary agreement for their participation. The core group also provides the leadership of the support group members with a budget projection, labor charge number, level of participation, and timeline.

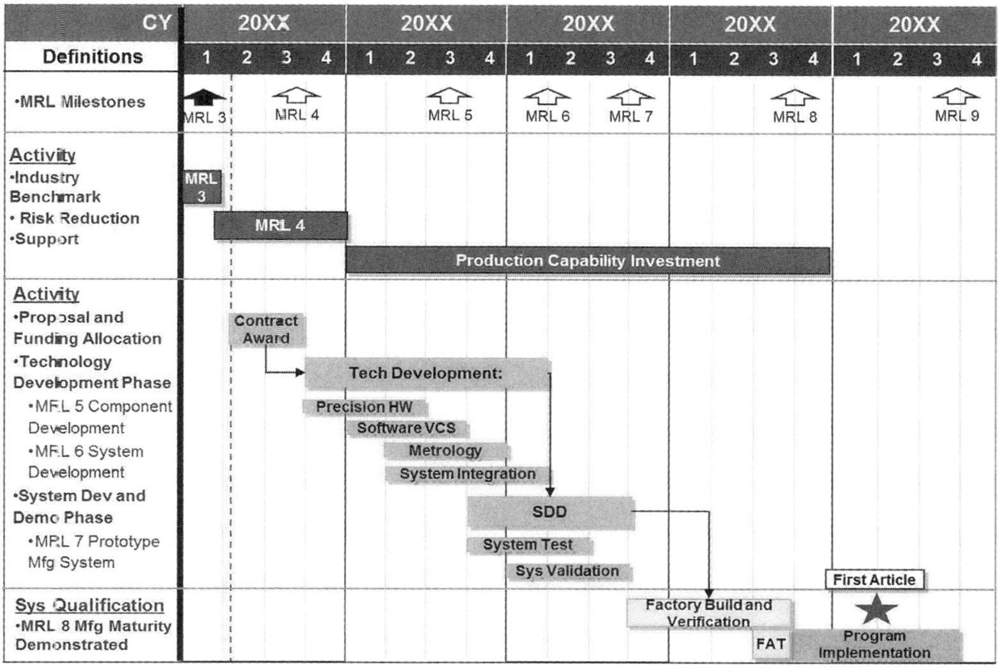

Fig. 9.1 Representative TRL/MRL template for automated D/C transition to production.

The TTP core group includes internal and external company members. The TTP core is inclusive of the machine designer, builder, and integrator.

While the formal establishment of the TTP team occurs on or just before the TRL/MRL milestone 4, the individuals who comprise the core of the team have been engaged independently from the beginning of the automation process. Their involvement and contribution include the automated system's preconcept, concept refinement, and concept approval as well as the affected manufacturing process identification.

9.6 TTP Document

TRL and MRL guidelines are extremely important for manufacturing risk identification and management. They must begin at the earliest stages of technology development, and continue vigorously throughout each stage of a program's life cycle. Matters of manufacturing readiness and producibility are as important to the successful development of a system as those of readiness and capabilities of the technologies intended for the system. And they are integral parts of the baseline development of the Transition to Production roadmap and TTP document.

However, the essential element to determine the success of the on-assembly automated D/C is the anticipated and agreed performance of the machine. All the elements and details of the TRL/MRL model can be met, but the outcome in quantity of holes drilled/countersunk, time to perform the tasks agreed, and at a quality level that meets standards must also be met. The best laid plans, well executed, that do not result in the number of holes committed at a speed and quality agreed, can and have resulted in difficulty and sometimes failure to get a signed acceptance of the machine by product delivery.

The feasibility study provides a target number of holes that are used as a rough order of magnitude (ROM) for estimating. The numbers entered into the TTP document must be absolute, attainable, and agreeable. All of the activities to bring an automated system on-line will be wasted if the outcome is compromised due to the machine's inability to attain (drill/countersink) the agreed number holes.

Ranking	Cost Center	Accessible Holes	Savings Factor	Baseline	Alternate	Delta Savings		Machine Cost	ROI
ONE	A	3000	0.7	24000	16800	7200		950000	≈132 Units
FOUR	B	1500	0.7	12000	8400	3600		1150000	≈319 Units
TWO	C	4500	0.7	36000	25200	10800		2400000	≈222 Units
THREE	D	1800	0.7	14400	10080	4320		1215000	≈281 Units
Total		10800	0.7	86400	60480	25920		5715000	≈220 Units
Column1	Column3	Column5	Column6	Column	Column	Column9	Column1	Column11	Column1
		Threshold 1200							
	E	1150							
	F	1000							
	G	900							
	H	850							

Fig. 9.2 A typical hole count matrix that is included in the TTP document.

Figure 9.2 shows a typical hole count matrix that is included in the TTP document. It details the ultimate commitment to the Periodic Interval Requirements (PIR) to support the product delivery's manufacturing plan. If all the effort has been performed and even a small quantity of holes are missed, the production plan must be changed to incorporate new planning to include tools, processes, training, personnel, and hours necessary to make up for the automated shortfall. This adds expense and complicates the manufacture of product. That is why a detailed and comprehensive restudy of potential holes must be performed and included as the stated objective in the TTP document. The documented commitment should include drill rate, quantity, and quality. This data is the final measure of success. Therefore, as a critical focus along the path to the production readiness review, hole count and efficiency should be the continuous benchmark for success. It should be the first agreeable, attainable measure of success that the TTP and Product Delivery Team documents and establishes as a major metric for success.

9.6.1 TTP Final Hole Count and Cost Benefits Analysis

The objective of the TTP final hole count and cost benefits analysis is to finalize the agreement of hole quantity and cost savings between the TTP and Product Delivery Team. In order to determine a final hole count and cost savings, estimating assumptions and methodology have to be defined. The assumptions include the candidate assembly analysis vs. existing hand processes that are replaced by automated D/C machines. Potential cycle time savings are estimated and added to include expected savings from reductions of hard tooling. The assumptions contain the number of spindles the machine will use. The machine life cycle cost is included. Labor standards are estimated using the company standard Cost Estimating (CE) model

The Affordability/Producibility team provides the analysis using three approaches:

1. Conservative approach based on an ROM estimate of 30%.
   a. Conservative analysis data is derived from:
      i. Improvement measured from baseline drilling methodology.
      ii. Cycle time improvements measured though implementation of robotic automation
      iii. Cycle time improvement is estimated at 30% of baseline.
      iv. Calculated cut-in effectivity.
      v. Estimated in future Total Annual (TA) dollars or euros.
   b. The conservative analysis results in an output of:
      i. Delta savings.
      ii. Payback in ship-sets.
2. Moderate approach based on an ROM estimate of 40%.
   a. The moderate approach uses the same stepped process with the exception of cycle time improvement, where 40% is used.
3. Very aggressive approach based on an ROM estimate of 50%.
   a. The aggressive approach uses the same stepped process with the exception of cycle time improvement, where 40% is used.

Cost savings analysis from each method is based on improvement measured from baseline two-step manual methodology (template drilling and manual countersinking); cycle time improvements measured though implementation of robotic automation and elimination of manual process; elimination of hard tooling required for the manual process; and estimates in projected TY dollars or euros of effectivity.

The analysis results in the nonrecurring tooling cost savings and the baseline improvement savings over the projected life of the program.

The conservative, moderate, and aggressive estimates are included in the TTP document as agreeable goals for each application of the automated D/C machine.

9.7 TTP Risk Identification and Assessment

Critical to the successful development and installation of automated D/C technology are risk assessment tools to identify, quantify, and classify risk. Risk assessment is the determination of quantitative or qualitative value of risk related to the development and installation of automated systems into an aerospace production line. Each auto-drill/countersinking scenario has unique risks associated with it. Therefore, each installation requires a fresh assessment of risk or recognized threat. The *quantitative risk assessment* requires calculations of the contributory components of risk *(R)*. They are the magnitude of the potential impact *(I)*, and the probability *(Ip)* that the impact will occur. Most risk assessment models express failure in terms of loss value. In the automated machine risk assessment, it is the weight assigned to the impact's effect on the system development and deployment. In the design, development, and application of complex auto-drill/countersinking systems, sophisticated risk assessments are made within the TTP and Product Delivery Teams when the risk concerns threats to life, environment, schedule, quality, machine functioning, or product. Once identified and ranked through the probability of impact assessment, a risk mitigation plan is put in place to reduce the risk to acceptable levels or completely eliminate it.

Risk assessment consists of an objective evaluation of risk. Assumptions and uncertainties are clearly identified, considered, and presented. The potential loss and probability of occurrence are difficult to measure. The chance of error in measuring these two concepts can occur but is reduced by participation of the team using their collective expertise and an Excel spreadsheet populated and weighted with a list of impact topics and converted to measureable ranking and outcomes. Risk with a large potential failure impact and a low probability of occurring is often treated differently from one with a low potential loss and a high likelihood of occurring. In theory, both are of nearly equal priority. The blind Excel assessment provides an objective outcome for risk assignment. The MRLs are assessed in multiple dimensions to populate the risk spreadsheet [9.6].

An important category of risk assessment that is outside the internal control of the TTP team and has the largest affect on for automation's transition to production is the technology and industrial base. These areas supply the machinery and the elements such as guidance systems, laser positioning systems, machine controller, tool bits, and other contributors to the function of the machine. All of these elements are necessary and must merge at critical milestones to contribute to the precision and performance necessary to meet the specified machine operating parameters.

Impacts to the completion and integration of hardware and software from the technology contributors can occur over the one- to two-year development, validation, and installation cycle. Many components are tied to a global supply chain that must be monitored to determine the continued timely and economic support of the project. Currency fluctuations, economic trends, technology challenges, and governmental regulations can impact the critical components' delivery. As part of the TTP plan and risk assessment, sensitivity analysis is performed to determine critical technology and provide alternate suppliers for backup.

9.7.1 Technology and Industrial Base

Identification of risks associated with the technology and industrial base is critical to the success of the Transition to Production of the selected automated D/C machine. Automated drilling machines are custom built to match the assembly cost center's configuration and meet its needs. There are very few suppliers who have sustained a viable business base to support the continued manufacture of these highly specialized machines. Some very reputable machine manufacturers have gone out of business at a critical point of the machine's construction, leaving the aircraft assembly facility scrambling to recover lost schedule and cost.

There have also been cases in which the aircraft assembler has returned to an automation designer/manufacturer of their machine for support, only to find the doors shut and business gone.

Compounding the financial risk associated with custom on-assembly automated D/C machines are the Intellectual Property issues that surround a piece of machinery built for a specific use. The airframe assembler has paid for its design and development, and therefore it feels compelled to capture the unique attributes of the machine in a patent and restrict access to the machine's special attributes by others outside the company. This approach has been and continues to be risky. The machines are of such specialized design, and the number of machines needed by any one manufacturer is very small. Restricting the machine manufacturer to selling only to the sponsoring airframe assembler affects the financial viability of the supplier. This, in turn, jeopardizes the airframe assembler's ability to support the machine.

Recently, there have been two approaches to mitigate the risk to the machine manufacturer's financial health of restricted revenue from single-source sales.

One approach is the strategic alliance approach that engages the machine supplier and makes them a partner for integration and sustained operation of the machine. The strategic alliance recognizes the supplier's skills for automation application to enhance the process. This relieves certain tasks (and stress) from internal departments that would have to provide resources (labor) to support the project. The supplier becomes engaged as a partner for success and is paid from the development and integration budget.

One solution of merit is the integration of the supplier as a strategically allied partner within the development, transition to production, and sustaining operations of the machine they supply.

The other approach is to free the supplier from the intellectual property binds that prohibit sale of similar-type machines to the buyer's competition. While this may seem counterintuitive to good business and competitive posture, the benefit of having a financially viable supplier/partner far outweighs the negative benefits. The supplier is grateful for your release to sell and make money, and your first application of the technology tends to provide an edge in its use.

The hybrid of these two types of approaches has evolved to include both. Sustaining the use of the automation is preferable to losing a supplier that is critical to continued use of the technology.

The other area that has changed in recent years is the implementation of direct supplier support of the drills and countersinks, preset in tool holders with spares and supplied to the line. Direct support of the line by supplying kits at point of use reduces the cost of maintaining a setup crib with stocks of inventory.

The evolution of strategic alliances, cost, and complexity of the automated drill/countersinking machines have led to the evaluation of buying holes instead of machines. This philosophy shifts the ownership and burden of developing, installing, operating, and sustaining the machine to the supplier. The per-hole cost is negotiated based on the same estimating system in Section 9.6.1, TTP Final Hole Count and Cost Benefits Analysis. By selling holes, the supplier has a vested interest in the machine performance, maintenance, and performance improvement. The machine is upgraded by the machine supplier to enhance performance because of the economic incentive to reduce D/C time.

Other multiple dimensions to populate the risk spreadsheet are:

- Design
    - Producibility program
    - Design maturity
- Cost and Funding
    - Production cost knowledge (cost modeling)
    - Cost analysis
    - Manufacturing investment budget
- Materials
    - Maturity
    - Availability
    - Supply chain management
    - Special handling

- Process capability and control
  - Modeling and simulation of production and process
  - Manufacturing process maturity
  - Process yields and rates
- Quality management, including supplier quality
- Manufacturing workforce (engineering and production)
- Facilities
  - Tooling, special test equipment, special inspection equipment
  - Facilities
- Manufacturing management
  - Manufacturing planning and scheduling
  - Materials planning

The cumulative outcomes of multiple MRL dimensions have a collected value expressed mathematically as E = pi (ph): the probability that the impact of any dimension will have on the project multiplied by the probability that the impact will happen, where E represents an event. Said another way, what is the probability that an impact *event* will happen, multiplied by the probability that it will affect the positive outcome of the project. The outcomes form a risk ranking from highest to lowest so resources can be allocated to mitigate the greatest risk. The evaluation criteria includes availability of alternatives that are viable substitutes should failure occur [9.7].

Ranking	Cost Center	Number Of Holes		Baseline	Alternate	Delta Savings	Machine Cost	ROI
ONE	A	3000	0.7	24000	16800	7200	950,000	≈132 Units
FOUR	B	1500	0.7	12000	8400	3600	1,150,000	≈319 Units
TWO	C	4500	0.7	36000	25200	10800	2,400,000	≈222 Units
THREE	D	1800	0.7	14400	10080	4320	1,215,000	≈281 Units
Total		10800	0.7	86400	60480	25920	5,715,000	≈220 Units

Fig. 9.3 Screen shot of an Excel spreadsheet showing the ranking system to calculate values.

Figure 9.3 is a screen shot of an Excel spreadsheet showing the ranking system to calculate values for the assessing risk. It is a blind ranking system that is averaged from the input of all the TTP and Product Delivery Team members. The members input the numbers, which are then multiplied by a risk contribution value in the Excel spreadsheet that is hidden from the team member.

The ranking focuses on the highest impact areas of risk that lead to an optimal allocation of resources, where harm to the project can be avoided [9.8].

9.8 Validation of System to Production Requirements: Exit Criteria

The final validation and transfer to production is the closure event that officially migrates responsibility to the production team. The signing (sign off) of the TTP document is the final step for factory acceptance of the automated D/C machine by Product Delivery. The TTP details, critical milestones, and timelines are included in a final document affirming acceptance of the responsibility to operate and maintain the automated system by production operations. The validation of the system is the satisfactory completion of the agreed tasks and milestones with a final passive test run or runs, as agreed. In addition, a single successful full run of the machine to perform all the tasks specified in the Requirements Document must be performed on "fly away" product. One hundred percent of the Requirements must be met prior to transition to production. The required exit criteria are the performance of the machine to the satisfaction of product delivery.

9.9 Summary

Transition to Production is the last step in the process to install automated on-assembly D/C of holes for fastener installation. The Transition to Production team integrates with the Product Delivery Team to ensure that a clear and concise understanding of the machine and its effects on the airframe assembly process are communicated. The medium for communicating the steps, outcomes, and risks to the Product Delivery Team is the Transition to Production document. The Transition to Production document is composed and populated with the steps, milestones, expected outcomes, risks, and risk mitigation plan for the final acceptance of the machine into the production process. The TTP document is a collaborative compilation of all the tasks necessary to ready the machine for production's independent use of the machine as an additional asset used to assemble the airframe. When signed by Product Delivery and the Production Operation's leadership, the document represents a formal acceptance of the machine into the production process. The risk mitigation plan and stepped process for integration into the production process uses the U.S. DoD established and industry accepted Technology Readiness Level and Manufacturing Readiness Level tools to build a detailed plan for progressive and verifiable progress. Detailed steps are defined that are addressed progressively through the levels until the completion of the ninth level, when the machine is transferred into the production process and operated, supported, and maintained by, or at the direction of, production operations. The determinants contained in the MRL tool are used to assess and determine critical risk technology or tasks that may negatively impact the progress or success of the implementation. Risk tools are used to evaluate the potential impact to the transfer to production project and also to determine the probability that the risk event will occur. The risk assessment is used to evaluate the greatest risk to the project and develop risk mitigation plans and alternative options for the risks. The risk probability analysis enables resources to be focused on the risk that would have the greatest impact for success and highest probability of occurring. The identified high-risk technology or activities are captured in a risk mitigation road map that corresponds to the master schedule for transition of the automated D/C machine into production. The risk mitigation schedule uses critical management tracking tools such as EVMS to maintain fidelity in tracking the risk mitigation plan. Tracking tools such as EVMS are also used to communicate and provide Production Operations leadership with visibility to observe and assess the progress of critical elements of the project.

The TTP document also finalizes the agreed outcome (results) at completion of the project. The number of holes the machine will access, drill, and countersink is part of the document and machine acceptance criteria. Also included is the efficiency and quality performance of the machine. At the end of the transition to production process, all the steps have been met to bring the machine online. As a final performance demonstration, the machine is "dry run" on the product at full specified speeds and feeds using a pen dot test. The rate of performance is validated to the TTP document. The dots are measured and assessed for positional accuracy using laser tracker technology. When the positions are confirmed and validated, Production Delivery accepts the machine by signing the document accepting responsibility, authority, and accountability for its performance, maintenance, and sustainability. Transition to Production is complete.

The transfer of RAA for the automated machine to the Product Delivery Team and production operations does not mean that the interaction with the core TTP team and its members is dissolved. Just as children are released to their own skills and abilities when they take that first bike ride sans training wheels and parent, the TTP team is available for support and counsel if there is uncertainty or struggles. The absence of the TTP team and its dissolution encourages the Product Delivery Team to stand alone with the new machine. But the core TTP team members should also be there to help if they are needed. It is a delicate balance.

References

9.1 Wheeler, D. and M. Ulsh, "Manufacturing Readiness Assessment for Fuel Cell Stacks and Systems for the Back-up Power and Material Handling Equipment Emerging Markets," Technical Report NREL/TP-560-45406, February 2010.

9.2 *Best Practices: Capturing Design and Manufacturing Knowledge Early Improves Acquisition Outcomes (GAO-02-701)*, July 2002.

9.3 "Interim rule, 76 FR 38050," Federal Register, June 29, 2011.

9.4 "Final rule, 76 FR 71645," Federal Register, November 18, 2011.

9.5 "Department of Defense Instruction 5000.02—Operation of the Defense Acquisition System," DOD, December 8, 2008.

9.6 *Manufacturing Readiness Level Deskbook*, DOD, May 2, 2011.

9.7 Glannon, Joseph W., *The Law of Torts* (3rd ed.), New York, NY: Aspen Publishers, 2005, pp. 73–74.

9.8 Flyvbjerg, Bent, "From Nobel Prize to Project Management: Getting Risks Right." *Project Management Journal*, Vol. 37, No. 3, August 2006, pp. 5–15.

Chapter 10
Training

Not every difficult and dangerous thing is suitable for training, but only that which is conducive to success in achieving the object of our effort.

—Epictetus

10.1 Introduction

At first glance it would seem that training for operating an automated D/C machine would primarily involve those individuals selected for the operation of the machine. However, due to the unique nature of the machines, their insertion into assembly lines to perform operations previously done by hand, and safety considerations, the extent of training runs through every discipline associated with production of the airplane. This includes part fabrication, engineering (all disciples including design), assembly operations, leadership, quality, and maintenance activities that are part of the design and manufacture process.

The insertion into the production stream of automated machinery disrupts every aspect of the part fabrication and assembly process. The decades of training and tribal knowledge that have evolved to high precision and craftsmanship are now challenged by machinery. Responses to challenges clearly understood by every person in the airplane's production stream are now invalidated. Compounding the change is the mixed application of hand and automated processes.

The training of people who have exclusively dealt with addressing challenges and processes for a hand-built airplane is paramount to the success of the automated system. For example, cost center managers are almost exclusively individuals who have displayed proficiency and understanding of the process used to build an airframe by hand. They have displayed the leadership, characteristics, and knowledge necessary to meet the challenges of the big three: cost, quality, and schedule. Realigning the workforce within the cost center the meet the challenges of the big three was a "talent" derived through understanding and realignment of labor. The automated D/C machines introduce consistency and complexity into the world of the cost center manager. They have not had the experience, training, or education to cope with fewer people to perform the same task. The machine is a threat to their experience and ability to see themselves as an expert at motivating people and orchestrating their labor to maximize output. As a result, the cost center manager is one of the critical individuals for training. They are taught the new options and actions necessary to make the new member of their team effective. By training the cost center manager as a proponent of the automation rather than an opponent of automation, the ground-level leadership propagates the successful integration of automation into the production stream.

There are many areas where training decisions have to be made and individuals taught the differences between assumed processes ingrained over their entire career for hand-build operations. Decisions must be made on questions such as, who do you train to operate the automated D/C machine? Is the person that best understands the assembly process, inclusive of drilling, countersinking, and fastener installation best suited to run the machine? Is it easier to train an assembly mechanic to run a machine or easier to train a machine operator to understand the assembly process?

Besides the shop floor and hands-on operation of the machine, design engineers have to be trained to include the attributes necessary for NC programming to include the machine commands that position the machine to D/C the holes. Prior to automated drilling, many of these holes were not included in computer models as point vectors, normal to the surface by engineering. Hand layout of master models and templates from log books identified hole size and spacing starting from a known position. Design engineers interacting with automated machinery for D/C have to be trained to understand what has to be included in the computer model for successful transfer to machine language.

This chapter will identify individuals and disciplines that must have the unique training that provides the understanding of how to integrate automated D/C into the assembly process. In addition, it will define what kind of training and the level of training and certification necessary for installation, operation, and sustainment of these complex machines in the formerly exclusive world of hand operations and processes.

10.2 Who to Train

Training is one of the first steps toward successfully integrating automation into the airframe assembly D/C process. Training begins during the evaluation phase when leadership is considering the decision to invest in automation. Training continues for selected employees and includes the entire journey all the way through to the automated system's decommissioning and disposal. Training is especially important if on-assembly automated D/C is being applied for the first time.

There is a difference between information, orientation, promotion, and training. All of them contribute to a successful automation experience. All of them could be considered training in preparation for observation or involvement in the installation and operation of the machinery.

Keeping people informed provides a communication medium that makes people more comfortable with change. Information can be conveyed informally in many ways including internal signage, news releases, internal announcements and postings, and informal information sessions for the interested.

Information is also used to formally convey the progress along the path to success to stakeholders. A certain level of training, or at least orientation, is needed for stakeholders. A clear understanding of the effect of automation, acquisition, transition to production, operation, maintenance, training requirements, and challenges that will arise is important. The stakeholders and leadership are better able to plan and understand the impact of automation so events and challenges that will occur do not take root as rumor or conjecture. Understanding the cost and challenges of automating reduces the stress associated with surprises derived from assumptions and preconceptions.

Training realigns the expertise of certain disciplines within the organization that will be engaged. For example, experts at programming an NC machine tool will in all likelihood NC program the automated D/C machine(s). The approach to programming parts on a fixed milling machine is quite different than programming an on-assembly automated D/C machine. The former removes material (milling and trimming) from a single part, primarily on a machine bed while being held rigidly in place by a tool. The latter uses a highly complex multipurpose end effector to drill holes using a subroutine in a collection of parts composed of varying material types and loosely held together. A programmer who assumes that their skill set is immediately transferrable without training invites failure and cost.

The amount of training that must take place is determined by the degree to which an individual interacts with and influences the operation of the machine.

10.2.1 Operator

The closest individual to the machine who influences its success is the operator.

As the first step to training, the operators must be chosen from either an existing group internally, or from externally available individuals. Hiring inside has advantages if the resources are available. It promotes the inclusion of the employees in the selection process and provides individuals familiar with the company, procedures, product, policies, and manufacturing processes.

The classification to select and draw candidates to train for automation is a point of discussion. There are two different opinions.

One approach is to train a machine operator to become an automated D/C machine operator. This approach starts from the premise that the machine operator is most adaptable to the operation of the machine because of their familiarity with the controller and the machine operation. The argument against this approach is that the complexities associated with the D/C of holes is vastly different from operating an NC milling machine. If selecting an individual from the internal pool of existing machine operators, then training must focus on the airframe assembly process. This includes those activities in and around the cost center as well as the nuances associated with the unique airframe assembly's D/C activities.

The other approach focuses on the mechanic in the cost center as the best candidate. The opinion is that the complexities of the cost center are much different than those of a machining center for part milling and trimming. The mechanic is better able to observe, detect, and prevent negative events that a machine operator would not recognize. The event could be recognized and stopped. Therefore, the training for the mechanic has to focus on the machine operation and setup.

The added advantage of choosing a mechanic from the assembly centers is their familiarity with their fellow mechanics. Comradery within cost centers is developed over years from serving a common purpose as a team. Selecting "one of our own" promotes the automation within the cost center and provides a sense of pride. The added benefit is that the operation of the machine has the added incentive of increase in pay. Within the cost center, the added incentive provides another avenue for advancement.

Some researchers today believe that a very real problem in today's automation is the HMI. A problem that is causing considerable uneasiness involves boredom, particularly when the machine is under the control of the computer program and the operator is monitoring the performance of the automation. Unfortunately, the training that the mechanic/operator received in this area has not been recognized and optimized. A training factor in the real world is the reality that if all of the available modes are covered in training, then the operators are responsible for demonstrating knowledge and competence in all modes during operation [10.1].

Monitoring the repetitive operation of an automated system has human ramifications. Monitoring repetitive motion in a robotic system can be a boring job.

In some cases, operators remove part of the automation and utilize the remaining features to infuse themselves as an active participant in the process.

Another problem and area of training focus involves the automated actions of the machine generating overconfidence and complacency, resulting in overreliance upon automation. The operator response to normal operations becomes routine, and the failed response happens when faults occur due to the operator's overreliance on the machine's correct response to negative or abnormal events. Operators tend to rely upon the automation to the point that the normal checks that are inherent in good manual operations are sometimes disregarded [10.2].

To overcome this problem, the training for automation engages the operator in the machine operation and success. Training focus should include involvement, information management, monitoring, prediction, cooperative management of systems, and intent.

10.2.1.1 Involvement

To be effective, the machine operators must be involved, whether they complete the task by controlling the machine directly or simply manage other human or machine resources. Highly skilled personnel do not monitor for the possibility of negative events that have a very low probability of occurrence. Maximizing the use of technology requires training for awareness and involvement in the automation process.

10.2.1.2 Information Management

To be involved, the human operator must be informed. The information available has to include all of the data that is necessary to keep the operator actively involved in the operation. This includes the information required to keep the machine operator fully informed regarding the state, progress, and intention of the system. Otherwise, the human operator cannot hope to be meaningfully involved. Training for the human operator includes training to determine the form, time, and manner in which information is presented.

10.2.1.3 Monitoring

The machine operator for automated D/C must be able to monitor the automated system. Automated systems are fallible, and operators are the last line of defense that can control, or in some cases prevent, a system failure.

Operators must be able to monitor the system effectively. Training to know how the system is planning to accomplish its task is essential to the effective operation of the system.

10.2.1.4 Prediction

Automated systems must be predictable. An automated system can only be monitored effectively if it is predictable. Machine operators must be trained for normal operation of each automated task as well as its behavior during any failure events, so they can make manual corrections or stop the automation failure from going further.

10.2.1.5 Cooperative Management

The automated systems must also be able to interact with commands from humans to determine the legitimacy of the command. And the operator must be able to monitor the automation as the result of expected actions resulting from valid commands. Training to recognize the human-machine interaction emphasizes that humans are fallible and that highly capable computers/controllers can malfunction. The training provides the association and relationship between the operator and the machine to anticipate all of the circumstances that might be encountered in a line operation. Performance of the machines and operators must be monitored by each other. For example, the controller should be able to send alerts when operators have made an error, and the operator should be trained to recognize when automation is making incorrect decisions, understand the result, and stop the process.

10.2.1.6 Intent

Each element of the system must have knowledge of the other's intent. Training to cross-monitor the human-machine interaction must include training for the operator to understand the intent of a machine action. The safety of the operation to product, machine, tools, and employees can only be effective if the operator knows the intent of the action. The intent principal requires training to understand the motion of the machine and the communication conveyed through the HMI. Figure 10.1 shows an advanced HMI for an automated D/C machine.

Machine designers of an on-assembly automated D/C machine for airframe assemblies recognize the probability of human- and machine-generated errors. Humans and machines working in an open environment of the modern assembly factory will encounter error as the result of their actions. It is wrong to assume that a perfect automated system exists to operate error free in a pristine environment. The existence of automation itself contributes to an environment in which an error can happen. Production operations cannot tolerate rare, inadvertent, or even random machine operational errors without controlling their operational consequences.

On occasion, change to the automation's operation such as enhancements or programming changes to the system drive the physical automation to action that causes mistakes; or the machine encounters a rare situation that the designer of the automated system failed to predict. The only resource that can deal with such an event is the human component. This makes the machine operator the ultimate system back-up to stop or minimize damage or injury. Training the operator to recognize these anomalies and what actions to take engages the operator and reduces incidents of boredom associated with operating automated machinery.

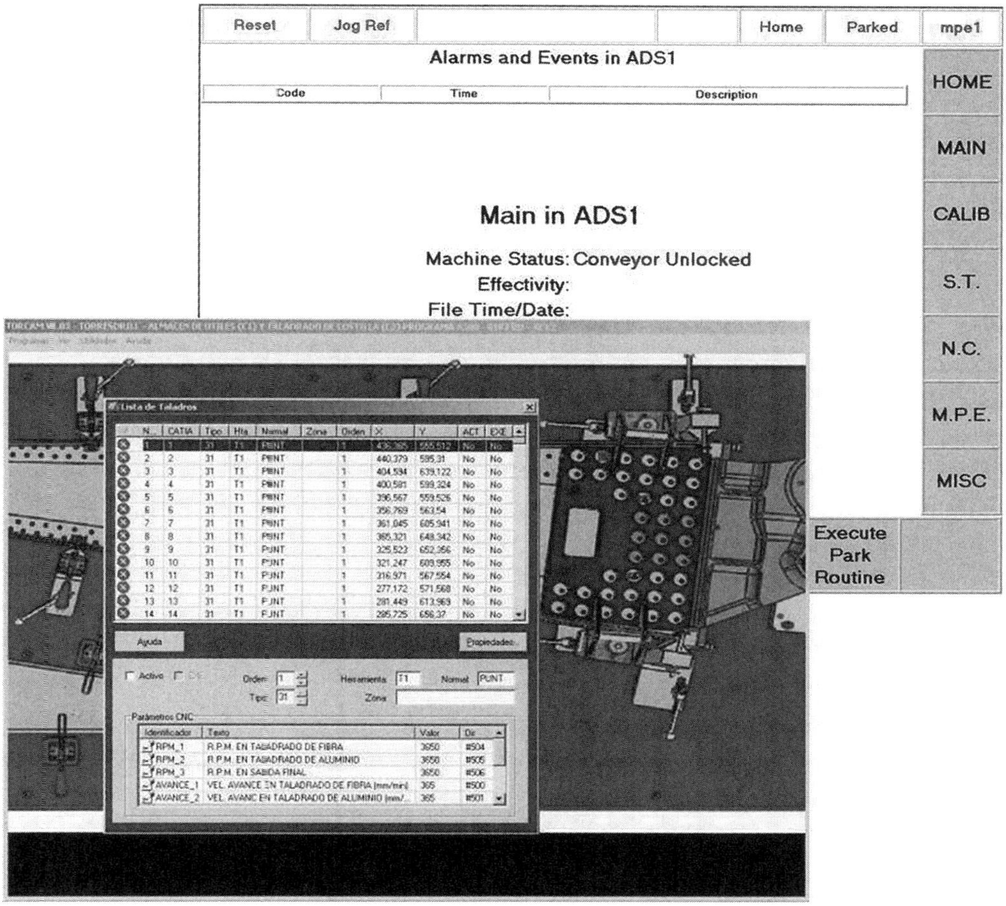

Fig. 10.1 Advanced HMI for an automated D/C machine.

10.2.1.7 Training to Deal with Errors

The automated system should be designed in such a way that it is made "error tolerant." Having the system error tolerant means that inadvertent errors are able to be detected and alleviated so that they are not critical. Simplifying the system allows unintentional errors to be detected and the appropriate action taken to correct the error. Modern HMIs have built-in action alerts with action-oriented option analysis [10.3]. They also have a communication medium so that alerts are aligned with knowledge experts. Alarm messages are sent to the appropriate knowledge expert for analysis and communication with the operator. The HMI also has the ability to affirm that actions taken by the operator are correct, and to provide communication via email, voice, and visual methods to reaffirm an operator or machine action is correct. This is a valuable tool for training the operator as they become familiar with the machine or to reacquaint them with infrequent events or machine motion.

Safety issues derived from operator error are important, and the advanced HMIs that can be linked to knowledge experts and maintenance personnel help to alleviate the operator's sense of isolation. Monitoring the actions of the operator remotely and providing input and communication to reassure the operator of their performance improves their confidence.

Training for the operator to monitor systems and performance of the automation are broken into three categories that are classified as systems, collision, and performance. All of the alerts and responses associated with these three categories are basic training requirements.

1. Systems include the electrical, hydraulic, mechanical, controller, and pneumatic systems that cause the automation to function.
2. Collision alerts are those that are emergency alerts notifying the operator to immediately stop the machine.
3. Operational alerts are those that have to do with the operations the machine is tasked to perform.

One of the great myths associated with automation is that it will reduce training requirements. Automation has created training requirements that add to the previous requirements. The skills and knowledge needed to take full advantage of automation must be added to the training curriculum. There is no doubt that automation makes D/C of holes safer and more efficient. It does not mean that today's operators do not need all the old skills and knowledge to adequately assess and respond to a machine's actions. Manual skills must be a part of any recurrent or transition training in addition to the emphasis given to the proper use of the automation.

There are many reasons for operator skill deterioration, including machine operating frequency and motivation. In most skills deterioration, the problem is lack of practice. For example, at the start of an airplane program, the frequency of operating the automated machinery may come only once or twice a year. The infrequent operation causes lapse in retention and prevents the best tool for learning, repetition. The way to help with the infrequency dilemma is early involvement of operators in the setup and validation of the NC programs. Another way is the development of simulations of the machine operating routines (including random alerts) to train and maintain efficiency in between actual D/C of flight hardware. Many suppliers of controllers will offer loan or lease of a comparable controller to load programs onto for the purpose of training and proficiency maintenance.

The human operator is and will remain ultimately responsible for the operation, efficiency, and safe operation of the machine. There are factories that are "lights out" when running automation. Given the expensive product and the exacting tolerances required for on-assembly D/C, this role will not change regardless of how advanced the automation may become. Automation systems are tools designed and used by human operators to complete tasks, and human operators are responsible for the operation of the automation systems. Solid training is the best insurance for success.

10.2.2 Design Engineer

Design engineers and structural engineers are at the front end of the process for successfully applying automation. CATIA V5 and other software packages are used by engineering to design and model airframes that are used by manufacturing. Manufacturing uses the computer models to translate the engineering to make tools and NC programs to make parts. The parts are combined to make subassemblies that become airframes.

In the hand assembly process, the tools use certain attributes of the engineering design model to convert to useable programs that can drive machines. Drill bonnets and templates are made from engineering models and dimensions. The templates' and some drill bonnets' construction uses spacing to determine hole positions. For example, engineering may define the position (called vector) of the first hole and the last hole in a run of holes, and then define the spacing in between the first and last hole. For example, the instruction might say drill 100, 0.250-inch-diameter holes between point A and point B. Figure 10.2 shows a typical example of how engineering defines many of the holes to construct tooling.

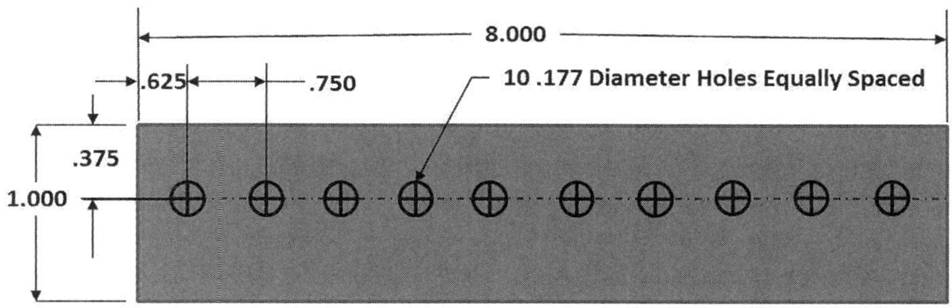

Fig. 10.2 Typical example of how engineering defines many of the holes to construct tooling.

In the hand process, a template with a run of holes is clamped to the skin of the airplane and the drill vector normalized to the surface of the airframe using bushings. The countersink uses a pilot to align the countersink to the hole. The face of the countersink cage sits flush to the skin to normalize the countersink to the surface. Figure 10.3 shows the application and proper use of a hand-applied countersinking cage.

If the multistep process is performed correctly, it results in a hole and countersink that is 90° to the surface of the airframe. The hand operation also relies on the mechanic to feel the torque and vibration of the hand drill motor to vary the speeds and pressure needed to successfully drill the hole.

Fig. 10.3 The application and proper use of a hand-applied countersinking cage.

The application of a machine to replace the hand process requires the inclusion of computer-generated information from engineering that is neither present nor required for the hand D/C method. The data must be included in the new model for automation to enable NC programmers to generate a program to position the machine at each hole, at the proper angle, and activate a subroutine for the stack of material types and thickness of the assembly.

The information is available to the engineer, who uses the information to generate specific data for each hole. The data for each hole includes hole number (identifier), hole size (diameter), hole vector (X, Y, and Z position), and angle of the hole (normal to surface; I, J, and K). The X, Y, Z and I, J, K coordinates are Cartesian coordinates used by the NC programmer to calculate the position and rotation of the machine for placing the drill tip in the proper position and orientation relative to the airframe. Figure 10.4 illustrates an example of a Cartesian coordinate system.

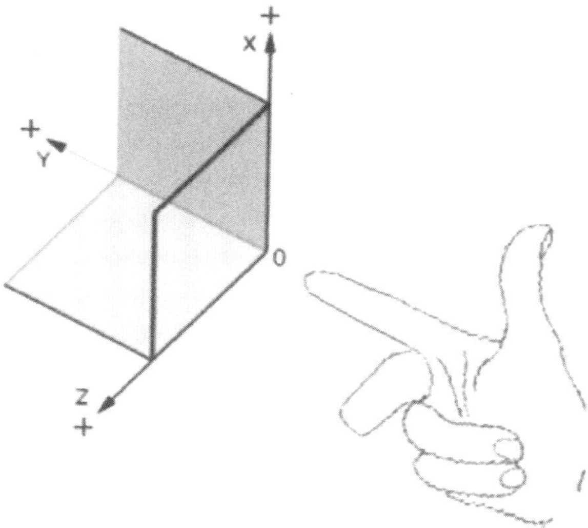

Fig. 10.4 Example of a Cartesian coordinate system.

The holes should also be identified as "approaching" the skin (tagged) from the outside of the airframe. Occasionally tagging the hole from the inside without discovery by the NC programmer causes the machine to drive through the airframe to drill a hole from the backside.

In addition, recent application of automated systems on older airframes such as the Northrop Grumman T-38 Talon wing has introduced the need for engineers to translate older paper drawings into 3-D engineering models. The T-38 was designed in the 1950s before computers were available for airframe designers. At the end of the production run, the paper drawings were scanned into Microfiche and stored. The re-winging of the T-38 to extend its service life presented a unique challenge—data had to be retrieved from a medium no longer used and the 2-D numeric information converted to a computer-generated engineering model for use by NC programming.

The engineering management and the airframe design engineers should be trained to provide the additional data, the type of data, and the format of the data needed by NC programming. While this adds expense to the design engineering effort, the alternative is for NC programming to retrieve an incomplete engineering model and create a usable model. NC programming's retrieval results in assumptions. The NC-generated 3-D model has to be sent to engineering for check and validation before release to generate an NC program. The back and forth effort creates additional expense to develop a 3-D model and an NC program.

Design engineers are under schedule pressures and budget constraints. Their primary responsibility is the engineering model for the airframe and its components. The specific development of hole placement has not been a priority. Therefore leadership and design engineer training for automation includes understanding the needs, expense, and schedule impact of automated drilling. The training includes identification, coordination, and cooperation with manufacturing engineering, NC programming, and the automation engineers.

Modern engineers have become "tube jockeys." They sit in a cubicle and design airplanes in a virtual world, and some have never stepped onto a production line. When their work is converted to work instructions for the mechanic, transformation from virtuality to actuality takes place. The mechanic is tasked to identify the anomalies that exist between the two worlds. They are also tasked with communicating the anomalies back through the process using action boards or other communication media. One of the mechanic's skill sets lies in the ability to catch disconnects between a virtual engineering world and the shop floor. While this frequently happens and engineering changes result, it also quite often does not. Pressure to build a good airplane to schedule interrupts a communication process that is meant to correct the virtual-real world dilemma. The mechanic has the ability to alter hole positions and mask an engineering error.

When automation is applied, the engineer is trained to enter the manufacturing world and become an active participant for identifying and correcting engineering error. Automation drills holes full size, and every action assumes correct location. If defects occur, they result in a costly chain of events that ripple back through many layers of bureaucracy to engineering. It is better to involve d esign engineers early in the process and train them to include the unique requirements needed to automate a section of the airplane.

10.2.3 Manufacturing Engineer

The manufacturing engineer defines all the elements of manufacturing that translate engineering design to reality. They perform the translation from engineering to reality by applying concepts, processes, tooling, technology, and other enabling activity to create a product at a defined rate, price, and quality. They also provide the communication medium to convey the translation through documents that include planning, work instructions, work orders, and tool orders. Anyone who has watched the programs, *How It Is Made* or *Modern Marvels* realizes that the magic is not in the product. It resides in the complexity of the manufacturing processes and application of technology that make the products that people desire at a price they can afford; at a quality level they expect; in quantities that enable wide distribution. All of this is made possible by the manufacturing engineer.

There are many fine colleges and universities that can provide multiple levels of education in the manufacturing engineering discipline. There are also universities such as Purdue University in the United States and Cranfield University in the United Kingdom that offer aerospace-specific programs. There are professional societies such as SME and SAE International that offer aerospace-specific training and certification.

Industry recognizes that education, training, and certification provide a solid generic foundation for a manufacturing engineer. They cannot provide finite training targeting the unique requirements of different products. For example, the specialized processes and technologies make it different to engineer the manufacturing of an automobile and the manufacturing of a television set. They share common generic manufacturing conditions such as the need to control dimensional tolerance, but the manufacturing methods are different. Within products there is also a difference in manufacturing engineering requirements. A door for an automobile is manufactured differently than an engine. Filling the gap (becoming industry ready) between a university-educated or society-certified manufacturing engineer and tailoring the knowledge to a specific product's needs requires time on the job. The more complex the product, the longer it takes to be industry-specific "ready." In addition, where multiple manufacturers produce products that compete with one another, competitive and proprietary manufacturing methods compound the specific requirements for a manufacturing engineer. A Ford method of manufacture may be different than a Volvo method of manufacture. When I began a training program at North American Aviation, I was told, "There is a right way, a wrong way, and a North American way. We will train you for the North American way."

In order to minimize the gap for manufacturing engineers and speed the path to industry readiness, industry uses training and orientation tools. They include targeting individuals in areas of expertise such as robotics for internship programs. University internship programs integrate people into the manufacturing culture for specific products so they can become familiar with processes and the company. They also gain valuable knowledge from experienced manufacturing engineers that introduce them to the nuances associated with successfully manufacturing a specific type of product.

Companies also engage in directed research programs that provide funding to universities. The funded research engages potential employees in projects directly related to product manufacture. Manufacturing engineers from the funding company provide guidance and counsel to the students during the research project. The relationships that develop help build teams and evaluate the potential for students engaged in the project to integrate into the company culture as an employee.

Regardless of the training and pre-employment engagement, nothing can prepare a new manufacturing engineer for entering the highly complex world of airframe assembly. Every airframe manufacturing engineer is nurtured by experienced manufacturing engineers who have been involved in building assembly lines. The tribal knowledge of how to assemble an airframe is passed from generation to generation of manufacturing engineer. And it has deep roots in the method to build airframes by hand.

Therefore, the internal training of existing manufacturing engineers includes the introduction to automation and its effects on the hand-built assembly process. It also includes the effects of islands of automation typically found in airframe assembly centers and how to build bridges that convey product in and out of the islands efficiently.

Introducing automation to a manufacturing engineer that is an expert at building airframes by hand in large, rigid, fixed tools is the first step in the training process. The training includes understanding the impact of motion on the assembly process and tooling considerations to facilitate its success.

Integrating automation into airframe manufacture and facilitating its successful interaction with existing hand processes is the second step of training. The automation changes the upstream and downstream assembly process as well as the cost center where it is applied. Rate, quality, and tooling consideration must be made to accommodate the changing environment of the automated cost center.

Engaging the machine builder and integrator early, identifying the manufacturing engineers to be trained, and reaming them together provides in-process exposure and training. Many of today's machine builders and integrators provide training programs to facilitate the introductory, intermediate, and advanced training of personnel.

An important area of training for the manufacturing engineer is in the area of cause and corrective action. When a defect occurs, the manufacturing engineer is called to evaluate the defect, determine the cause, and recommend corrective action. The evaluation most often is narrowed to people, processes, or tools. Once the evaluation is complete, corrective action is determined and carried out. For example, a determination might be, *mechanic error due to setting the countersink depth too deep, causing a deep countersink.* Corrective action might be, *additional training required for mechanic on setting depth of a countersink.*

When automation is introduced and a defect occurs, there is a level of complexity that is unfamiliar to the airframe (hand-built) manufacturing engineer that they must evaluate, determine cause, and recommend corrective action. Training must focus on the area of machine fault analysis and operational conditions that can contribute to machine error. The training to expand the awareness of events outside the hand process is essential to find, correct, and reactivate the machine.

10.2.4 Quality Assurance

Embedded in the hand process are the quality engineers and inspectors that certify the product conformance to manufacturing and product specification. The hand airframe assembly process engages the quality personnel in an interpersonal relationship with the mechanic, in which the inspector is called and responds by evaluating work performed by a specific mechanic. The reputations of the mechanics are built by their ability to meet the standards certified by the quality inspector.

When machines are introduced, a level of anxiety is created because of the change of manufacturing process. A small group of holes and edge distance drilled by one person is inspected in the hand process. Large groups of holes and edge distance require inspection in the machine process. The need to keep the machine running uninterrupted through a large quantity of holes and hole sizes requires longer dedicated inspection times. In addition, the need to return the asset back to productive work provides a level of urgency to the task. Mechanics can perform other tasks while waiting for the inspector to evaluate the holes. The machine cannot.

The definition of hole quality does not change. Size, concentricity, position, edge distance, and other criteria are the same whether done by hand or machine. Training needs to focus on the change in how to approach the inspection operation to optimize the machine performance. Part of the integration involves alerts that the machine is running and predictive notice of machine completion so the inspector can be on-site at the completion of an operation that requires quality conformance signoff.

10.2.5 NC Programmers

A technology that enhances the efficiency of programming an end effector for automated D/C is Position Based Drilling and Countersinking (PBDC). PBDC is designed to make NC programming faster and easier with automated programming. PBDC can provide higher quality as a result of fewer errors and can deliver greater standardization through the use of proven and preferred processes and resources. The PBDC concept uses software that can recognize a location and its relationship to the engineering definition for material type, thickness, and countersink depth. The software determines the best or preferred speeds-and-feeds routine for each process to D/C the hole, then generates the proper routine for each operation at the correct location—all automatically. Automated D/C programming is created using a wide range of alternative machining sequences and storing each one in a template; each one effectively represents another variant of the D/C process. The material type, hole size, pecking routine if needed, and depth are all examples of factors included in the D/C template package.

Creating and using this type of rapid programming for automated D/C is performed by programmers trained in its construction and application. Many automation software suppliers such as Applied CAx offer training courses to bring NC programmers up to speed on programming for automated systems.

10.2.6 Cost Center Managers

One of the critical challenges when automation is applied into a cost center where hand operations are performed is resistance from the cost center manager. The cost center manager is the direct supervisor of the mechanics of the cost center where subassemblies are assembled. The manager also serves as the communication interface between elements of the manufacturing process that govern the success of the cost center including tool control, production control, quality, and master scheduling. The fluid motion of the entire production line into and out of the cost centers depends on and is the responsibility of the cost center manager.

Cost center managers typically are promoted from the mechanics that work in the cost center or in one like it. For example, a mechanic working on a product tail assembly in one cost center would be eligible to manage another product's tail assembly. As mechanics for a particular type of assembly, they become intimate with the assembly, its construction, and the personnel that build it. Within the cost center, they are the ultimate authority. Because each puzzle piece of a production line must fit precisely, in the right sequence, and at the right cost, everyone is subservient to the cost center manager's request for action. They rule the cost center. Their performance is assessed by what comes out, when it comes out, and how much it will cost before it comes out of their cost center. Their skill lies in their ability to manage people, realign resources, recognize issues, prevent disruption, and anticipate the unforeseen.

When automation is presented as an addition to their cost center, it creates extreme anxiety. At first glance, everything about the introduction of automation into the cost center is a threat to their livelihood and position. Automation comes with an automatic budget reduction to the cost center, based on its projected savings. Human resources are reduced, providing fewer options to the manager to shift resources where they are needed. A new technology threatens the security of the manager as an expert for assembling the airframe.

All of these anxieties can be turned to positive training to acquaint the cost center manager with their new friend. Training in this arena focuses on the enhanced performance and consistency of the machine to facilitate better numbers for the cost center. Training also focuses on past success and how the machine acts and interacts within the cost center. Cost center managers are trained on the basic machine operation, how it functions, and how they manage the new technology. Their involvement and introduction to these aspects of automation reduces anxiety and improves their confidence to manage the new technology as another tool in their tool box.

10.2.7 Tooling

In the airframe assembly tooling world, tooling engineers and tool designers are familiar with large, robust assembly tools with very little mechanized motion. Moving product from tool to tool most often involves cranes. Therefore, the introduction of motion by automated D/C machines is not an area of expertise of the tool engineer or designer.

They are experts in the design of static tools made for the hand process. Motion introduces a level of complexity that requires advanced software and training in simulation, finite element analysis, safety systems, and machine dynamics. The extent to which the company or factory has embraced automation should determine the depth of the training required by the resident tool engineers and tool designers.

If there is only to be one or two machines, then basic training is essential. However, advanced design and integration should be left to the automation integration specialists to work with the tool engineers and designers to derive the best solutions.

If the automation is to be extensive, then a core group should receive in-depth training and work alongside the supplier integration team. When the integration team is finished, the core group becomes the company's resident experts for expansion or upgrades.

10.3 Summary

Training employees for automation takes place after orientation. Training for automation will enhance the skills, capabilities, and knowledge of employees for performing all the tasks necessary for installing and operating automated D/C technology. This is especially important when introducing automation into processes and activities previously done exclusively by hand.

Training for automation includes three approaches for success.

1. New candidates for training should be identified early to begin integrating them into the culture of the company and product manufacturing processes. The training familiarizes them with the organization, mission, vision, rules, regulations, and working conditions of the company.
2. Training is made for existing employees to acquaint them with the automation and how it operates within the organization. Purchasing automated D/C technology changes the way an airframe is produced and how it interacts with surrounding processes and people. Training facilitates an understanding of how the new equipment and processes work.
3. Training is conducted to assure employees of the positive benefits that result from automation, such as promotion and career growth, as well as easing the physical effort of their job.

Not all the departments or individuals that should be trained are identified in this chapter. At a minimum, the critical departments to select individuals for training should be the operator of the automation, design engineer, manufacturing engineer, NC programmers, quality assurance personnel, cost center manager, and tool engineers and designers.

The decisions of who should be trained, and how much training the individuals should receive, are decided and tempered by many factors. Budget is one consideration. However, one thing is clear. Not training is more expensive than the cost of training adequately.

References

10.1 Orlady, H. and L. Orlady, *Human Factors in Multi-Crew Flight Operations*, Brookfield: Ashgate Publishing Ltd., 1999.

10.2 Bullen, George N., Machining tool operator training system, U.S. patent number: 6033226, Filing date: May 15, 1997, Issue date: Mar. 7, 2000.

10.3 Management study guide, pathway to success, http://www.managementstudyguide.com/training-of-employees.htm, Retrieved June 4, 2012.

Chapter 11
Maintenance

Another flaw in the human character is that everybody wants to build and nobody wants to do the maintenance.

—Kurt Vonnegut Jr.

11.1 Introduction

Maintenance and support for D/C automated machinery differs from the maintenance and support for standard machinery. D/C automated machines are highly specialized and integrated into the assembly line. They require support and maintenance that is unique and critical to the continuance of the production flow. Standard machines and customized D/C automated machines differ in that standard machines perform repetitive tasks that can be performed by other machines in the factory inventory when the standard machine is taken offline for maintenance, calibration, or repair. If the other machines in the factory are at 100% utilization, the standard machinery is common enough to offload the temporary requirement to other parts producers to maintain production continuity by manufacturing and supplying the needed parts. The common nature of the standard-type machine means that spares inventory, maintenance mechanics, and calibration technicians have a number of machines that are related and in common. Therefore, spare parts inventories, training, and calibration raise the maintenance, repair, and calibration of the standard machines to a routine part of the factory operation.

Because automated D/C machines are custom built and integrated into the production assembly flow, they require higher-level training, attention, and response. When an automated machine is transitioned into the production line, the machine's calibration must be aligned with the maintenance and calibration rhythm of the production line. Taking the machine offline for calibration outside the assembly line operations rhythm disrupts the line flow because no other machine can take up the slack. The same applies to damage and repair. The automated D/C machine operates in the highly populated environment of the assembly line, where other tasks and functions that surround the machine are dependent on the tasks it performs. A damaged or malfunctioning machine stops the line flow before and after the assembly center where the machine resides. A damaged or malfunctioning machine needs emergency response because there is not inventory to draw from to maintain the continuity of the production line. The machine is performing an inline task that cannot be replicated offline to sustain line flow.

Also, spare parts are unique to the machine. Therefore, inventories of common parts for multiple machines cannot be kept in stock.

Further compounding the maintenance and repair issues attributable to the automated D/C machines is their custom construction. They were custom designed and built and therefore reside in very limited numbers or in many cases as a single unit on the production floor. Maintenance and calibration personnel may have been involved or trained to do minor maintenance, but the lack of repetitive application of their skills reduces or challenges their proficiency, compounding the ability of the factory to bring the machines back online.

This chapter will address all of these issues and provide the necessary actions and response to maintain and calibrate the highly complex automated D/C machines that are integrated into the assembly production line. It will also describe and provide tools to determine critical spares and skills lists, maintenance and calibration options, emergency response options, and effective preventative measures to reduce machine down time.

11.2 Complexity—New Normal

Maintenance for automation inserted into an airframe assembly line changes the normal and established culture of maintenance. Large, fixed assembly tools are made to rigidly hold pieces of assemblies together while they are being fastened. An aircraft assembly line has dozens and sometimes hundreds of the major assembly tools installed in factories. Because of their size and mass, they are assumed to be robust enough to perform their intended task uninterrupted for a year. Once a year the factory shuts down, and maintenance and calibration are performed on major assembly tools. In between, their functional operation is assumed to be in conformance if no defects in the parts are determined to be caused by the tooling. There are exceptions to the annual maintenance and calibration other than tooling-caused defects: examples such as damage where a crane carrying an object has impacted the tool, or a detail of the tool has broken or been damaged. These are minor events, specific to a single tool, that do not interrupt the assembly line. Night shift or day shift response teams from the tool maintenance department repairs the tool while in service or while it is inactive and not needed. Figure 11.1 shows an aircraft assembly line populated with large, robust assembly tools.

Fig. 11.1 Aircraft assembly line populated with large, robust assembly tools.

Rarely will an assembly line be shut down entirely for damage assessment, interim calibration, and maintenance. One example is the earthquake rule in California. Major assembly tools are leveled to the horizon around their base to within 0.002 in or 0.05 mm. The leveling of the assembly tool is performed to establish a datum plane that is used for the tool reference plane and to position all the details of the assembly tool. Tweaking the baseline horizontal datum plane can affect the location of the tooling details that position the part. An earthquake above 6.5 magnitude on the Richter scale can shake the tooling and change this critical datum plane and alter the position of components in an assembly. The assembly line is shut down after a 6.5 event, and all of the tools are checked and re-leveled where needed. Earthquakes of this magnitude are rare.

The maintenance, calibration, and repair of major assembly tools in the world of the hand-assembled airframe are annual or reactive. Because of the nature of the hand assembly of airframes, humans are engaged in every activity. Tooling-caused defects are observed and caught, usually before they escape to the next operation, and often in process. The mechanic notices that something is not right, notifies the appropriate person, an evaluation is performed, and corrective action is done to repair or recalibrate the tool.

When automated D/C technology is integrated into an assembly line, a new normal is established. The new normal is exponentially more complex to maintain, and exception or annual maintenance, calibration, and repair is no longer adequate or preferred.

There is an old saying that if one person can do a job in thirty hours, then thirty people can do the job in one hour. The story illustrates several differences between humans and automation. Eyes on the activity provide in-process evaluation of the activity. If there is a damaged tool or the mechanic's drill bit breaks or chips, they stop and call for maintenance or replace the drill before continuing. Automated D/C machines without PdM capabilities do not have eyes on the activity and do not possess intellect to evaluate, stop, and correct a potentially negative condition. When thirty people are applied to the same area to speed up the operation, there are thirty sets of eyes on the activity. Each evaluating, stopping (if needed), and correcting a potentially negative event within their area of responsibility. Machines are optimized to the maximum output. They move from hole to hole with the assumption that each action and all the contributors that affect correct action are aligned to make a good hole and countersink. The machine does not possess the ability to determine if all the contributors in its environment are correct. Therefore, any machine that operates without eyes or intellect to observe, stop, evaluate, and correct a potentially negative event can produce defects at a very rapid rate before they are detected. The eyes and intellect of a machine are the technologies, software, and alarms used by the machine to perform PdM.

In the earthquake example, a mechanic would stop drilling and hold on until the shaking stopped. The act of sensing and stopping prevents vibration transferring into the drill operation that might cause a defect [11.1]. The act of stopping work until the shaking subsides prevents multiple defects. Machine devices such as vibration sensors stop the automated drill/countersink/fastener installation operation of the machine in Stuart, Florida when trains pass the facility. Otherwise, the machine would keep drilling, impervious to the defects it was causing. Another type of technology, installed on major assembly fixtures in El Segundo, California, uses jacks, sensors, inclinometers, and a programmable logic controller to maintain the level of the fixtures, or re-level them after an event such as an earthquake [11.2].

These systems provide PdM, or condition-based maintenance to evaluate the condition of equipment by performing periodic or continuous (online) equipment condition monitoring. Condition-based monitoring (CM) allows maintenance to be scheduled, or other actions to be taken, to avoid the consequences of failure before the failure occurs. Nevertheless, a deviation from a reference value such as vibration must occur to identify impending defects or damage. PdM does not predict failure.

Once a condition begins that can cause a defect, the failure process has already commenced and CM systems can only measure the deterioration effect of the condition. Intervention in the early stages of deterioration is more cost effective than allowing the machine to fail. Condition monitoring provides the benefit of detecting conditions that would shorten normal lifespan. They can then be addressed before repeated failures occur [11.3].

The greatest threat to the machine during its operational life is an impact causing unforeseen or unscheduled maintenance, repair, and recalibration of the machine.

11.3 Impact Prevention

The unique position of on-assembly D/C brings it in proximity to people, objects, structures, tooling, and misplaced objects such as clamps. The movement in and around complex structures in small enclosures also creates potential for collision and impact that damage the end effector. Figure 11.2 shows an end effector pressure foot in an enclosed space.

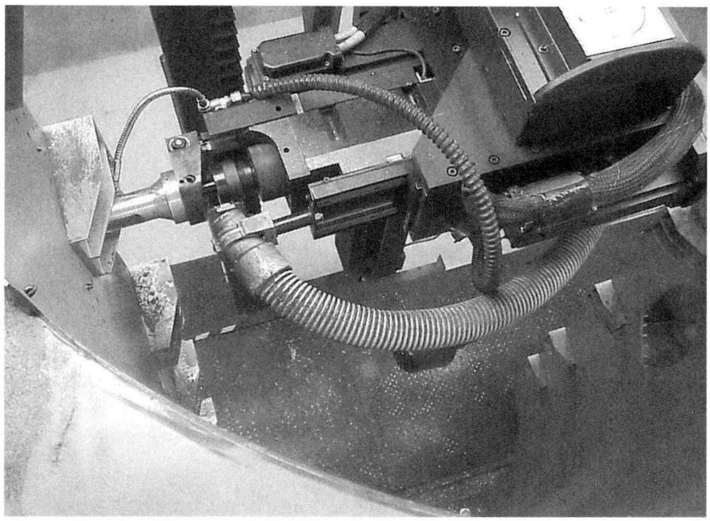

Fig. 11.2 End effector pressure foot in an enclosed space.

The end effector is the most costly part of the machine other than the machine structure. It contains all the technology that makes possible the D/C operation.

Therefore, an end effector collision avoidance system or automated collision-potential shutdown system integrated with the machining operating software is an important PdM process monitoring attribute. Figure 11.3 shows a damaged end effector resulting from unforeseen impact with an object inadvertently placed in the way of the machine motion.

Fig. 11.3 Damaged end effector resulting from unforeseen impact.

11.4 Predictive Maintenance

PdM or condition-based maintenance attempts to evaluate the condition of equipment by performing periodic or continuous (online) equipment condition monitoring. The ultimate goal of PdM is to perform maintenance at a scheduled point in time when the maintenance activity is most cost-effective and before the equipment loses performance within a threshold. This is in contrast to time- or operation-based maintenance, where maintenance is performed whether the equipment needs it or not. Time-based maintenance is labor intensive, ineffective in identifying problems that develop between scheduled inspections, and is not cost effective.

Automated D/C machine process-measurements that can be gauged for PdM are speed/rpm, head pressure, weight, and position. Ultrasonic sensors that are integrated with condition monitoring software are available that can produce alarms or e-mail notifications when thresholds are exceeded. The alarms trigger calls for maintenance, repair, or calibration activity.

11.4.1 Maintenance Integration

Condition monitoring systems and smart sensors are becoming increasingly sophisticated, less expensive, and more prevalent for use in the plant environment. Data can be gathered, analyzed and presented in an actionable format during the process, in-situ. However, the diversity of condition monitoring technologies, software, and processes can cause confusion. Simplification and sound requirements definition will facilitate the ability to construct a system to compile data for the operator to view from the HMI showing asset health and control functions.

To expedite predictive maintenance tasks, condition monitoring suppliers are developing interfaces to enterprise asset management systems so that work orders, inventory requests, and associated processes can be activated automatically.

While many companies will spend enormous amounts of time and money on tools, equipment, parts, and materials, they will not focus on developing the foundation of a good maintenance organization—the maintenance system. Using predictive techniques without an effective maintenance system in place only optimizes your reactive maintenance program. It will result in marginal savings and less-than-anticipated payback. Predictive maintenance is good, but you must have the other programs in place to support it [11.4].

Reliability-centered maintenance (RCM) uses predictive maintenance (PdM) techniques in addition to traditional preventive measures. When properly implemented, RCM provides companies with a tool for achieving lowest asset Net Present Costs (NPC) for different levels of performance and risk [11.5].

Often overlooked is how to transfer PdM data to a Computerized Maintenance Management System (CMMS). Development of protocols so that the equipment condition data is sent to the CMMS object to trigger maintenance planning, execution, and reporting is important. Otherwise the PdM solution is of limited value.

To evaluate equipment condition, predictive maintenance uses nondestructive testing technologies such as infrared, acoustic (partial discharge and airborne ultrasonic), corona detection, vibration analysis, sound level measurements, oil analysis, and other specific online tests. New methods in this area utilize measurements on the actual equipment in combination with measurement of process performance to trigger maintenance conditions. This is primarily available in Collaborative Process Automation Systems (CPAS). Site measurements are often supported by wireless sensor networks (WSN) to reduce the wiring cost.

Technologies that can be integrated into the collaborative process for automated D/C include holographic imagery, vibration, acoustics, and infrared.

11.4.1.1 Digital Holographic Imagery

Advancement in technology for process performance evaluation that can be linked through WSN for Collaborative Process Automation System analysis is digital holographic image sensors. Digital holographic image sensors measure the impact, deflection, and result of on-assembly D/C. The system provides an all-inclusive three-dimensional (3-D) image of the process as it happens. The *should-be* and *as-performed* conditions for hole location and size are constantly monitored and compared to the engineering/manufacturing specifications. The data is managed by software that contains the metrics for hole position and size.

A classical hologram is a quantitative, dense record of an entire observable object. This optical record is formed instantly by the interference between two laser light beams of a single wavelength, one reflected from the object and the other used as a reference wave. The interference pattern of the light beams allows the 3-D shape of the object to be created, with depth resolution finer than the laser light's wavelength.

A digital hologram is recorded on a digital camera. Using several wavelengths of laser illumination, the digital hologram integrates multiple classical holograms into an entire observable object record that quantitatively covers an optical 3-D depth range and tolerance of surface characteristics. The results are far beyond those measurable with a single classical hologram while still equaling or exceeding the fine depth resolution achievable with a single wavelength hologram for the same surface type. Figure 11.4 shows the output from a digital hologram image.

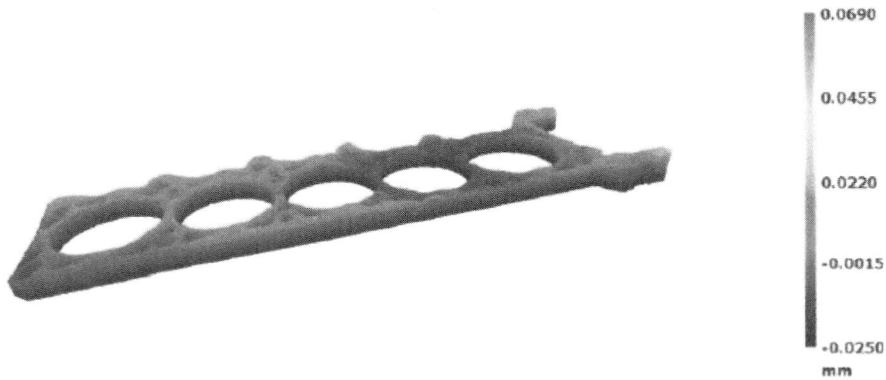

Fig. 11.4 Visual output from a digital hologram image.

The system measures precision surface dimensions to the micron level, measures hole sizes and locations, and detects a variety of surface dimensional defects. Figure 11.5 shows the sensor head unit of the digital holographic high-definition Shapix surface detective.

11.4.1.2 Vibration

Vibration analysis is most productive on high-speed rotating equipment such as automated D/C machines. Vibration analysis allows the user to evaluate the condition of equipment where end effector vibration or external vibration could cause defects. The latest generation of vibration analyzers is sensitive and has capabilities and automated functions that can effectively monitor and discriminate between vibration types. Many units display the full vibration spectrum of all axes simultaneously [11.6].

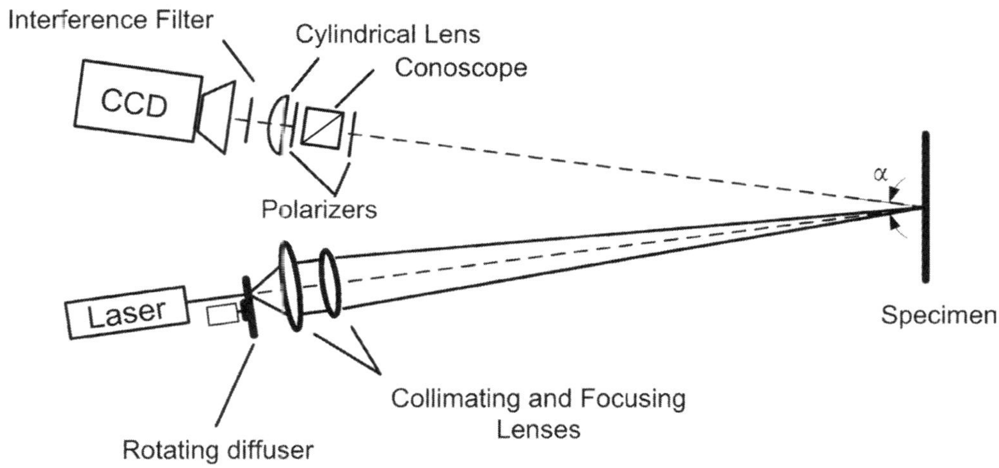

Fig. 11.5 Sensor head unit of a digital holographic high-definition surface detective.

11.4.1.3 Acoustics
Acoustical analysis can be done on a sonic or ultrasonic level. Ultrasonic techniques for condition monitoring make it possible to "hear" friction and stress in rotating machinery. The acoustic signature can be used to predict deterioration earlier than conventional techniques. Ultrasonic technology is sensitive to high-frequency sounds that are inaudible to the human ear and distinguishes them from lower-frequency sounds and mechanical vibration. Machine friction and stress waves produce distinctive sounds in the upper ultrasonic range. Changes in these friction and stress waves can suggest deteriorating conditions much earlier than other technologies such as vibration. With proper ultrasonic measurement and analysis, it is possible to differentiate normal wear from abnormal wear, physical damage, imbalance conditions, and lubrication problems based on a direct relationship between asset and operating conditions.

Sonic monitoring equipment is less expensive, but it also has fewer uses than ultrasonic technologies. Sonic technology is useful only on mechanical equipment, while ultrasonic equipment can detect electrical problems and is more flexible and reliable in detecting mechanical problems [11.7].

11.4.1.4 Infrared
Infrared monitoring and analysis has the widest range of application (from high- to low-speed equipment). It can be effective for spotting both mechanical and electrical failures. Infrared used for oil wear particle analysis determines the mechanical condition of machine components that are lubricated. Through wear particle analysis, composition of the solid material present can be analyzed and evaluated for particle type, size, concentration, distribution, and morphology.

This is not new technology and has been used successfully to determine mechanical wear. Helicopter maintenance personnel have used oil wear analyzers to predict the need for maintenance on transmissions.

11.5 Prognostic Health Monitoring

The integration of all of the PdM technology is combined and is used as an in-process Prognostic Health Monitoring (PHM) system to monitor and manage the automated system's health.

PHM systems monitor system health, provide early detection of faults, identify failure modes, point out failure precursors, detect degradation, determine remaining useful life, and recommend maintenance responses [11.7]. An automated D/C machine's long life cycle ensures the users' satisfaction and low probability of failure. Industry views success of the system to include availability. Need of higher operational availability and product liability due to system failures drive the need to estimate and predict the system's performance while in operation. PHM enables maximum operational availability, minimum periodic inspections, a low number of spares, maximum usage life, accurate part-life tracking, and no false alarms. PHM systems determine application conditions by sensing, recording, and interpreting the environmental, operational, and performance-related parameters that are indicative of a system's health [11.8].

Prognostics refer to the ability to determine operational availability, as well as to accurately determine the remaining useful life (RUL) of a system. The built-in test (BIT) and self-test abilities in a system were early attempts at providing diagnostic capabilities that were incorporated into a system's structure [11.9]. Applicability of PHM capabilities was limited to the failure definition embedded at the system's manufacturing stage. Due to more recently developments in sensor and data analysis capabilities, the implementation of data-driven diagnostic systems that can adapt to new failure definitions is possible.

An automated system's health assessment should include information on in-service uses, operations, and environmental conditions to enable prediction of failures and provide warnings in advance of catastrophic failure. The PHM processes sensory information by diagnostic and predictive algorithms. The algorithms help in determining the extent of deviation or degradation of a product from its expected normal operating condition. The objective is to : (1) warn of impending failures; (2) reduce unscheduled maintenance, extend maintenance durations, and increase system availability by needed actions when required; (3) reduce life-cycle cost of equipment by decreasing downtime and inventory; and (4) improve and assist in design and logistic support for systems in service and in the design stage [11.10]. Other basic requirements included in the algorithm are the ability to ensure that the collected data are accurate, consume fewer resources, build system history, and reject noise due to the measurement or operational environment.

11.6 Maintenance

The maintenance categories of a machine that fill out the PdM arena are autonomous, breakdown, improvement, periodic, improvement, and certification [11.11].

Preventive maintenance has been more popular in principle than in practice. Keeping equipment well maintained to extend its expected life and avoid future repair costs is an agreeable concept. The economic relationship between preventive maintenance and its RCI has been cloudy. Operators trying to gain preventive maintenance (PM) budgets from leadership for their maintenance portfolios present obvious arguments. The equipment will perform better. Equipment life will be extended. Repair costs will fall. Downtime will be reduced. Internal customer satisfaction will increase. The machine manufacturer says we need to do it.

Although leadership acknowledges these issues, a more convincing argument for preventive maintenance requires a scientific approach to demonstrate that PM generates a solid rate of return in risk mitigation and asset protection. Evidence does suggest that preventive maintenance is valuable. The economic value of preventive maintenance is difficult to determine. No specific statistical methods exist [11.12].

That is why PdM using in-process methods and technology purchased as part the machine can reduce cost. Participatory maintenance such as autonomous maintenance of the machine can help reduce costs. Each piece of automated D/C machinery is unique and operates in its own unique environment. Size, material, and other factors contribute to the cost and frequency of maintenance.

However, there is no doubt that the operation of a machine that performs thousands of operations on a piece of product that has a collective investment in the tens or even hundreds of thousands of dollars should have a detailed and comprehensive approach to maintenance. The malfunction of the machine can and has cost companies using them much more in product repair or rework, where a minimal investment in preventative maintenance beyond PdM would have prevented the loss.

11.6.1 Autonomous

Autonomous maintenance is performed by the operator at the beginning and end of operations to validate that the system is ready to perform its function. The maintenance requirements include visually inspecting the machine and making sure no damage has occurred before (idle time) or after the operation. It includes filling lubricant and coolant reservoirs as necessary and validating that screws and attachments are tight. The operator performs the same function that a pilot performs when they preflight an airplane. Both are trained to recognize damage or loose connections to critical components and parts that may affect the machines performance. In certain cases, they are trained to perform certain minor maintenance themselves. The auto-machine operator uses a written check list to walk through and evaluate the machine before operation.

During the operation, the machine operator maintains operation activity logs that note observations that may need attention. These are provided to the cost center manager, who passes along observed requirements to the manufacturing engineer or maintenance personnel.

In addition, they are trained to perform cleaning and removal of grit and debris that would inhibit the smooth or continued action of the machine. The cleaning and visual inspections are critical to the operation of the machine. Many issues have been detected and prevented by attentive operators that reported unusual wear or damage before the machine is put into operation.

11.6.2 Breakdown

Breakdown maintenance is performed as the result of machine failure. Machine breakdown is treated as an emergency "red call" alert. This is because a machine failure can result in loss of productivity or even damage to the product.

11.6.3 Improvement

Improvements are upgrades and enhancements to the machine. They include components that will measurably increase productivity or reliability of the machine. The improvements are derived from the machine manufacturer's research and from performance logs that are used as feedback to enhance the hardware design, software upgrades, or NC program revisions.

11.6.4 Periodic Maintenance

Periodic maintenance is time based and is coordinated with the maintenance rhythm of the factory. It is performed to check the machine as well as its coordination with the environment in which it is integrated. This includes tooling and parts.

11.6.5 Certification

Certification includes the accuracy calibration and systems check for continued acceptable performance. The calibration for accuracy is performed using laser tracker technology at the tool tip of the automated D/C machine to validate its continued performance. Calibration and certification is performed whenever there is concern derived from observed performance or on a timed-based system.

11.7 Buying Maintenance

Company leadership has to decide to what extent they want to maintain a trained team of maintenance personnel and spares for highly specialized machinery installed in limited numbers.

Buying and maintaining the critical spares (drives, end effector components, etc.) to support the machinery on-site provides immediate access for external and internal maintenance teams. The best maintenance support can come from the supplier. They have the expertise and equipment to perform periodic maintenance, improvement, certifications, and some breakdown repairs. The negotiated purchase of the initial machine includes recommended spares and may include a maintenance contract to perform these activities. Although breakdown contracts have been given to suppliers, they most often include timing arrangements in which a guaranteed time to arrive was designated as a line item in the contract. The period from call to arrive time for a red call alert is generally 24 hours.

Breakdown and autonomous maintenance should remain onsite. A breakdown is a critical event that draws the attention of the entire product delivery team. The shutdown is visible and may affect the product. Therefore, rapid response teams within the factory need to be trained to respond immediately to determine the extent of damage and call the machine integrator and supplier if they are not the same entity. Damage to the product is the first priority for damage assessment and communication to leadership. Assessment of damage to the machine by the response team needs to include possible parts that the machine supplier may need to bring and a clear communication of the machine's operation up to the event. This includes interviews with the operator or others that may have witnessed the event.

11.8 Troubleshooting

The process of troubleshooting computer-controlled manufacturing (CCM) systems presents difficulty for the expert maintenance personnel. Acquiring and using information to diagnose faults based on collecting background information, testing hypotheses, making repair attempts, and generating beliefs about causes of failures complicates the process [11.13].

One of the most common calls to maintenance and automation engineers is the complaint that the machine will not start. I mention this because as strange as it may seem, a common cause is the accidental or purposeful depression of the emergency stop button or obstruction of the safety systems. In the first instance, pulling out the emergency stop restarts the machine. In the second instance, a machine walk around will discover the obstruction. Removing it will restart the machine motion. Before removing blockage from in front of a safety stop, the machine must be put on stop or pause. Removing the obstruction in some machines will restart the motion. If someone is in the way, injury can result.

Many times, the installation and integration of automation onto an assembly line for airframes is the first or second experience with automated D/C a factory may have. The new kid on the block is always the first suspect when trouble shooting an on-assembly automated D/C machine. People tend to respond to trouble with a "what has changed?" approach to evaluation. Because of this, many obvious issues that could have been resolved by simply looking at the tooling, part placement, or other environmental issues around the machine have been overlooked. When troubleshooting an on-line automated D/C machine, it is important to evaluate the problem from a holistic approach that incorporates the entire operating environment. This includes tooling, parts, work-stands, operator, or other activities in the cost center.

11.9 Summary

Purchasing a new piece of machinery comes with euphoria and headaches. Pride of ownership and anticipated performance leads to a sense of satisfaction once the headache of installing and integrating a complex piece of automation has been accomplished. The headaches can and do return if adequate protections, including maintenance, are not included when installing automation. On-assembly automated D/C has a unique position as an integrated operating system. Its operation is coordinated and dependent on the reliability of tooling, processes, activities, and operations within its environment. Maintaining the automated system must be holistic and must consider the critical nature of its integration into the total process of control.

Impact prevention and predictive maintenance use technologies that enable the system to communicate and control events. These events are critical because they can harm the machine, parts, or tooling. Some of the technologies available to use are holographic imagery for in-process quality assurance and vibration sensing to determine machine wear and external environmental conditions.

PHM integrates all of the monitoring devices into a coordinated system to continuously monitor, maintain, predict, communicate, and improve the health of the machine to extend its life cycle.

Active maintenance includes autonomous, breakdown, improvement, periodic, and certification. The regular and continuous "eyes-on" of the operator performing cleaning, inspection, and control observation duties is the first line of defense against a machine's faults and break downs.

Regular calibration and periodic inspection by expert maintenance personnel ensures the machine's satisfactory performance and extends its life cycle.

Buying maintenance and breakdown support is an economic decision driven by the number and complexity of the systems installed at a factory. The expense to maintain certified automation maintenance employees, equipment, and spares is high. Many automation users buy maintenance from the machine integrator. If purchasing a maintenance agreement is the choice, then a time standard for emergency response must be a part of the contract. Even though a maintenance contract is in place, the user should, at a minimum, maintain critical spares and a trained crew to respond to emergency machine events such as crashes.

Troubleshooting crashes and other critical "red call" alerts should be approached with the entire system in mind. Any one of the elements contributing to the coordinated operations within the cost center can cause a negative critical event. Simple solutions such as inadvertent obstructions should be evaluated first. The first step of troubleshooting is to evaluate the damage to the part and the machine. The second step is to interview the witnesses to the event and then clearly communicate the extent of damage and the potential cause to leadership and repair personnel.

The cost of maintenance purchased as part of the machine or performed by trained personnel is expensive. Exact figures for an ROI for a given level of maintenance do not exist. Each machine and factory operates in its own unique environment, using machines in a variety of ways. Some models have averaged the expense as 15% of the purchase cost of the machine. The same models have estimated that minimizing maintenance expense costs the user 30% of the cost of the machine. This includes cost from shortened life, breakdowns, and diminished performance.

References

11.1 Mills, Simon R. W., *Vibration Monitoring and Analysis Handbook—(INST397).* The British Institute of Non-Destructive Testing. ISBN [[Special:BookSources/978-0-903132-39-7|978-0-903132-39-7]], 2010.

11.2 Bullen, George N., Self leveling independently programmable system, Patent number: 5587900, Filing date: May 23, 1995, Issue date: Dec 24, 1996.

11.3 ISO, ISO 17359:2011, Condition monitoring and diagnostics of machines—General guidelines, The International Organization for Standardization (ISO), 2011.

11.4 Pond, Mark, "5 Pitfalls to preventative/predictive maintenance techniques," Marshall Institute, maintenance Technology, http://www.marshallinstitute.com/default.asp?Page=Maintenance_Resources&Area=Articles&ARTID=Pitfallspdm, Retrieved June 3, 2012.

11.5 Mather, D., *The value of RCM Plant Services*, 2008.

11.6 Kennedy, S., "New tools for PdM," Plant Services, http://www.plantservices.com/articles/2006/072.html, Retrieved June 2, 2012.

11.7 Bereiter, Susan R., "Difficulties in troubleshooting automated manufacturing systems," *International Journal of Industrial Ergonomics*, Volume 5, Issue 1, January 1990, pp. 91–104.

11.8 Pecht, M, *Prognostics and Health Management of Electronics*, New York: Wiley Interscience, 2008.

11.9 Gao, R.X. and A. Suryavanshi, "BIT for Intelligent System Design and Condition Monitoring," IEEE Trans. on Instr. & Meas. 2002, 51(5): 1061–1067.

11.10 Vichare, N., P. Rodgers, V. Eveloy, and M. Pecht, "In-Situ Temperature Measurement of Notebook Computer—A Case Study in Health and Usage Monitoring of Electronics," IEEE Trans. on Device and Materials Reliability 2004, 4(4): 658–663.

11.11 "Total Productive Maintenance Overview 150," http://www.toolingu.com/class_class_desc.aspx?class_ID=%20900150, Retrieved June 6, 2012.

11.12 "Determining the economic value of preventative maintenance," http://www.pmmi.org/files/MS/certified/newsletters/PreventiveMaintenance.pdf, Retrieved June 5, 2012.

11.13 Bereiter, Susan R., "Difficulties in troubleshooting automated manufacturing systems," *International Journal of Industrial Ergonomics,* Volume 5, Issue 1, January 1990, pp. 91–104.

Chapter 12
Social Impact of Automation

My greatest challenge has been to change the mindset of people. Mindsets play strange tricks on us. We see things the way our minds have instructed our eyes to see.

—Muhammad Yunus

12.1 Introduction

During the last two decades, automation has become omnipresent in airframe assembly, and its advancement and rapid expansion has replaced processes and operations performed by humans. Automation's continued development, implementation, and daily use in the assembly of airframes have a profound impact on the traditional relationship between humans and machines. Their increasing use has strongly altered the organization, professional, and personal view of technical systems and how they interact and affect the daily manufacturing operations of the corporation [12.1].

Behavioral changes as the result of automating hand processes and operations have become apparent as machines and humans interact to achieve new balances between human and machine contributions to the production of airframes. The influences that contribute to the changing relationship between humans and automated machinery justify the consideration of the new vision for addressing the human-machine interaction.

The real or perceived social impact of automating a process has a profound effect on its success or failure. In many cases the perceived social impact has a greater effect than many of the real social impacts directly attributable to automating a process.

Understanding the core activities necessary for successful application of advanced automation technologies into the social environment is necessary to maximize their full benefit for humans and organizations. Maximizing the full benefit of automation for humans and organizations requires enlarging the traditional model of automated machinery, control technology, and social sciences into a hybrid model for holistically coping with the increased complexity of integrating human behavior with advanced technologies. The hybrid model for manufacturing includes the integrated socio-technical systems necessary to facilitate the incorporation, integration, and operation of highly technical machinery into processes previously performed by humans. The complexity of the human-machine relationship requires a continuously developing operational model that maintains an adequate balance between "pushing" automated technology into manufacturing while addressing its suitable use by the organization.

For example, more and more automation is surrounding assembly airframe mechanics in their work environment as well as their social environment. The pervasive effects of automation in their personal and professional life can cause controversy and influence the successful acceptance of highly automated systems to perform manual work. Their sense of the overwhelming intrusion by automation into every aspect of their life causes pushback against further change. This hidden social impact directly affects the expansion of automated systems.

This chapter will address the relationship between humans and the automated technologies that encompass all issues and elements of the social impact of automation to the assembly of airframes.

Included will be some of the latest applied methods and approaches that are helping to integrate automation into the human domain. Successful methods and approaches emerging are: Pacifying the Human Capability-Machine Experience; Collaborating Human Skills and Machine Intelligence; Balancing Automation and Human Processes; Addressing the Perception of Ubiquitous Automation; Safe, Secure, and Ethical Automation; and Management of Automated Systems.

12.2 Pacifying the Human Capability-Machine Experience

Each time the option to automate a hand operation is explored, the social impact of the human capability-machine experience becomes apparent. The pushback against automating any hand operation is experienced by the automation application engineer at all levels of production operations, from top floor to shop floor.

The best example of this can be described by personal experience. During my first application of on-assembly automated D/C in California (the first application on a military airframe), I experienced "pushback" by many individuals who displayed less than accepting attitudes towards automating the process. The pushback came in the form of covert and overt resistance. The automation was being installed on a new assembly line where during installation disparaging remarks shouted from the aisle were heard daily. The perplexing part was that the taunts and shouts came from members of leadership who had lost the vote to install the automation. During design, acquisition, test, installation, and transition to production, I had attributed the pushback as individual acts rather than a collective source of resistance that has become recognized as another step for consideration when addressing installation of automation into a hand operation.

The importance of addressing the social impact of pacifying the human capability-machine experience became apparent in Florida. After the successful installation of the automated piece of equipment in California, I was sent to study the possibility and potential of automating an airframe built at our company in Florida. The head of operations at the factory welcomed the team and sent us to an office that had been set aside for us located adjacent to the factory. We left the office and went to the factory to look at the airframe assemblies and to speak with the mechanics. There are three sources available to obtain initial information about an airframe: performance records, build logs, and the mechanic. The best information comes from the airframe mechanics, because if you can gain their confidence, they will tell how the airframe is built and where the traps and pitfalls are that are not conveyed in the other sources of information. They live with the airframe and know it.

When we entered the factory, we experienced momentary euphoria. The tooling was open and all the holes were exposed for access of machinery. The tooling was robust and could easily be modified to accept the application of machinery. There were long, straight rows of tens of thousands of holes the same size. They were small holes that were drilled in thin stacks of aluminum located to a very "forgiving" positional tolerance.

When I approached the mechanic in a cost center containing a fuselage section and introduced myself as a fellow employee that was here to study the process, I was greeted with, "Are you the guy that is here to install that robo-s__t?" My answer was that I was here to see if there was a way to make his job easier. I followed with the question, "How long have you been working in this cost center?" His answer was, "About 12 years?" Clearly he knew this section of the fuselage. "Have you been drilling these holes all that time?" I asked. "Yup." He answered. "About how many holes would you say are in this section of the fuselage?" I enquired. Stepping close to me he looked into my eyes and answered, "Hell son! I just drill them! I don't count them!" He then turned and walked away. This single experience personified the entire experience at the Florida factory. The combined resistance from top floor to shop floor to automating any process became evident over the successive weeks. To successfully automate any cost center at this factory would depend on the social acceptance of the new process and not on the technical applicability or monetary viability of the automation application.

This example is one of the many that have been experienced by automation engineers as automation begins to transform the proud traditions of manufacturing from hand built to machine built product. Automating a process without including, addressing, and pacifying the social impact of the human capability-machine experience throughout the production operations organization can negatively affect its application.

12.2.1 The Social Concept of Operations

When I became aware of the importance of addressing the social impact of automating a process or processes in a factory, I began to look for material to develop a standard set of procedures to incorporate into a social concept of operations. The procedures and S-CONOPS would be developed and integrated into the feasibility and implementation documents as first steps necessary to guide the determination and facilitate a more favorable and cooperative environment to obtain objective data. The processes definition and S-CONOPS would also reduce (pacify) tensions and anxiety of the workforce and leadership derived from using technology to replace human effort, skill, and knowledge to perform a process at lower cost. It would emphasize using technology as a fail-safe mechanism to monitor and increase certainty and control over production and the organizational functions [12.2].

As part of the process and steps contained in the S-CONOPS, accentuation of the positive attributes would be described and explained. They include reduction in repetitive work and tedious procedures, an increase in skill level, higher job satisfaction, an increase in the variety of tasks, and greater flexibility.

The potential negative effect of automation would also be addressed. They include possible degradation of the quality of working life, decline in interpersonal communication, increase of employee stress, depersonalization, boredom, lower job-satisfaction, loss of control over the pace of one's work and organizational functions, and lower self-esteem and staff morale.

Faced with a dearth of research in this area, a questionnaire was developed to use for interviewing the production operations organization personnel to determine their familiarity with automation and attitudes toward its effect on their work and personal life. The interview questions result in data that can be used to ascertain areas of concern to be addressed before any consideration of automating a hand process begins. The questions are meant to determine the degree of acceptability and devised to investigate the perceived effects on personnel at each factory where automation might be installed.

The questions concentrate on how an integrated, automated drill and D/C system is perceived to affect quality of working life, job satisfaction, client relations, self-esteem, morale, the pace of work, and the control of organizational functions. The questions are also meant to determine other socio-cultural aspects of automation such as which individuals and staff would be involved in the implementation of the system, staff training, and attitudes of staff toward the system.

To study the potential impact of the new system on production operations personnel, employees are interviewed and asked to answer a questionnaire. Both procedures are carried out twice more during the feasibility and cost benefits analysis study to maintain contact and mitigate concerns throughout process.

By asking questions about the workforce's perceptions regarding automation and its impact on their lives, data can be gathered to address the concerns and therefore reduce (pacify) their anxiety. Reduced anxiety eases the effort to obtain accurate data and information necessary to carry out the feasibility and cost benefits analysis.

In addition, data obtained from the questionnaire also identifies pockets of resistance where specific attention can be applied in the form of training, orientation, involvement in the process (inclusion), and briefings to address the specific concerns of an individual, group, or organization. In most cases, attention to addressing the concerns of individuals, groups, or organization reduces uncertainty. It is uncertainty that leads to anxiety and resistance to automation.

The questionnaire provides the option of anonymity to the responder, creating a more candid response to questions, resulting in better data. However, the distribution and timing of the questionnaire distribution allows some specific identification of where concerns reside. For example, if the questionnaires have limited distribution to a group or department, then the responses can be directly attributed to that group or department. This provides enough anonymity for candid response to the questions while providing enough specific identification of groups or departments that need to have their concerns addressed. Individual bar codes at the top of the questionnaire are keyed to specific groups so simultaneous distribution can occur while isolating and collecting data segregated by groups of individuals or organizations.

The questionnaires and the collected data are broken down into seven categories. The categories are quality of working life, job satisfaction, internal relations, self-esteem, morale, pace of work, and control of organizational functions.

The categories are divided so that they can provide a modular tool for the user to choose the most appropriate questions for a selected group. However, valuable data can be obtained by viewing the perceptions and opinions of departments and individuals other than those that the questions are meant to solicit a response.

The questionnaire serves other purposes. It provides a medium through which the respondent is led to evaluate their own perceptions and beliefs regarding automation. The act of seeing their concerns addressed, having to evaluate the answers to questions before answering, and seeing a list of questions reflective of leadership's desire to value their input can often soften the most ardent opponent of automation.

Another purpose of the questionnaire is to provide information that can be used to develop a potential automation team. The responses can reveal potential candidates.

At the beginning of the questionnaire, respondents are asked a set of multiple choice qualifying questions. How would you describe your work arrangement in your main job? How long have you worked in your present job for your current employer? In your main job, are you salaried, paid by the hour, or what? Figure 12.1 shows a generic set of qualifying questions and multiple choice answers.

Taxonomy of Job Stressors and Sample Items

Construct (Job Level)	Sample Items
Job demands	My job requires a great deal of concentration.
Job control	I have a lot of say about what happens on my job.
Meaningful work	Do you feel that the work you do is important?
Perceptions of risk	The safety conditions where I work are good.
Predictability of work	To what extent do unexpected events occur on your job?
Responsibility for Others	How much responsibility do you have for the morale of others?
Role Demands	I receive incompatible requests from two or more people.
Utilization of Skills	At work, how pleased are you with the way your abilities are used?

Fig. 12.1 Generic set of qualifying questions and multiple choice answers.

The questions provide the survey respondent with a selection between five answers. They are asked to select one of the answers by checking the box corresponding to Highly Agree, Agree, Disagree, or Highly Disagree. At the end they are asked to write a short sentence or two describing their overall attitude toward the automation of the D/C process.

A standard set of questions is used to baseline the respondent's current quality of working life, job satisfaction, internal relations, self-esteem, morale, pace of work, and control of organizational functions. The baseline is then compared and contrasted to responses to questions pertaining to their perceptions about how the introduction of automated systems will affect their current quality of working life, job satisfaction, internal relations, self-esteem, morale, pace of work, and control of organizational functions. Figure 12.2 shows a sample graph of one possible outcome to the compilation of data from a survey.

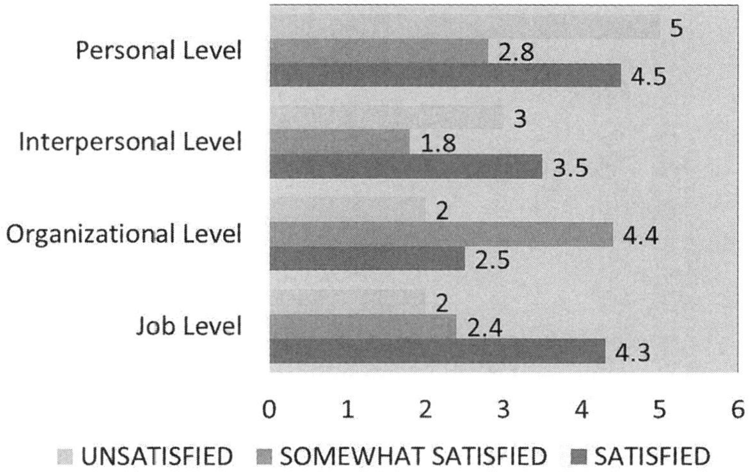

Fig. 12.2 Sample graph of one possible outcome to the compilation of data from a survey.

12.2.1.1 Quality of Working Life

Automation has often been accused of affecting the quality of working life [12.3] because it effectively allows for the deskilling of work. Tasks are broken down to their most basic level or replaced by the machine altogether, so that staff need little initiative or skills to carry them out. The job becomes very narrow, with very little variety, autonomy, creativity, etc.

This section's questions address the concerns individuals and organizations have about the perceived change to their quality of work life. They address quality of work-life related factors. These factors include working conditions, general well-being, home-work interface, stress at work, and control at work.

- Working conditions questions help indicate the extent to which an employee is satisfied with the fundamental resources, working conditions, and security necessary to do their job effectively and the impact they feel automation will make to the current quality of that environment. This includes aspects of the work environment such as noise and temperature, shift patterns and working hours, pay, tools and equipment, safety, and security. A negative perception of automation can have a significantly adverse effect on your overall outcome of automation evaluations and implementation if not addressed.
- General well-being questions help determine the current and perceived impact of change resulting from automation that the employee will experience. They are meant to assess the extent to which the employee feels good. The sense of general well-being may be more or less independent of the employee's work situation. General well-being both influences, and is influenced by, work. It reflects psychological well-being and general physical health aspects.

- Home-work interface determines how much the employee thinks the organization understands and tries to help with pressures outside of work. The home-work interface is related to work life balance and is about having a measure of control over when, where, and how you work. These questions help determine the extent the production operations personnel perceive that automation will improve or negatively affect their lives at home and at work.
- Stress at work is increased by any change in the routine. These questions look at the extent to which employees may see work pressures and demands made by the application of automated D/C as acceptable and not excessive or "stressful." Work pressures and demands can be a positive of aspect of the automation change experience, providing challenge and stimulation, but where seen as excessive and beyond our ability to cope, we are likely to feel overloaded and stressed.
- Control at work questions determine how far the employee feels they are involved in decisions that will affect automation design, utilization, and process application. Control at work (involvement in the process) reflects the level to which the employees feel they can exercise what they consider to be an appropriate level of control within their work environment. The perception of control might be linked to various aspects of work, including the opportunity to contribute to the process of decision making that affects their work tasks.

SAMPLE QUESTIONS: I get to do a number of different things on my job; On my job, I know exactly what is expected of me; Introducing automation will help by improving the quality of my work, and expanding the tasks I perform as part of my job.

The response to these questions would be Highly Agree, Agree, Disagree, and Highly Disagree.

12.2.1.2 Job Satisfaction

Job satisfaction is the perception that one's job fulfills or allows fulfillment of one's desires, expectations, and needs.

Job satisfaction questions help determine the degree an individual sees automation as increasing or diminishing their satisfaction for the work they perform. How they answer questions relates to whether they feel the automation workplace will provide them with an improved environment at work to generate a sense of achievement, high self-esteem, fulfillment of potential, etc.

Job satisfaction is reflected by questions asking how satisfied people feel about their work and is influenced by clarity of goals and role ambiguity, appraisal, recognition and reward, personal development, career benefits, and enhancement and training needs.

SAMPLE QUESTIONS: My job requires that I keep learning new things; My job lets me use my skills and abilities; The chances for promotion are good; I am given a lot of freedom to decide how to do my own work; Automating the D/C of holes will increase my ability to use my skills and abilities.

12.2.1.3 Internal Relations

Internal relations questions are meant determine the concerns of individuals that automation will diminish their human interaction with others. Some of the concerns include the need to concentrate on the screen or machine motion, operator not being allowed by management to help with anything more than general directional queries, lack of privilege level, lack of mobility, and system breakdowns.

It also determines the concerns of the supervisors that will be held accountable for the new automated processes' ability to interact with the other assets (human-machine interaction) of the cost center effectively. The questions also address the concerns that arise from a new piece of machinery and the supervisor's ability to manage it effectively and efficiently.

SAMPLE QUESTIONS: The people I work with can be relied on when I need help; In general, relations in my work place between management and employees is good; What I earn on my job in comparison to others doing the same type of work is fair; My supervisor is helpful to me in getting the job done; My overall interaction with my colleagues, supervisor, and friends will be enhanced by the introduction of automated D/C.

12.2.1.4 Self-Esteem

Self-esteem is the favorable opinion or approval of one's self that leads to feelings of self-confidence; worth; strength; capability; and of being useful, valuable, and necessary to the organization.

The questions in this section are meant to determine how the operators feel about loss of autonomy and control over their work.

SAMPLE QUESTIONS: The chances for promotion are good; I am proud to be working for my employer; At the place where I work, I am treated with respect; Automation will enhance the value I provide to others and myself.

12.2.1.5 Morale

Morale is a pervading sense of cheerfulness, confidence, excitement, etc. within the work environment. Staff morale seems to be a function of management policy, the quality of working life, and job satisfaction. Therefore, it will clearly vary in different work situations. Morale is measured by asking staff about perceived changes in the working environment that they believe would be effected through the application of automated D/C machines.

SAMPLE QUESTIONS: I am very satisfied with my job; The work I perform is stressful; During the past 12 months, I had trouble going to sleep or staying asleep; Automation will help me reduce the stress I feel at work and provide a better environment for me to work.

12.2.1.6 Pace of Work

These questions ascertain the individual and departmental concerns that the line flow and individual tasks previously understood will be increased by automation. The pace or rhythm of the production line is clearly understood and can be changed within and outside the cost center. The installation of automated systems introduces an inflexible system into a dynamic and fluid system of manufacture.

SAMPLE QUESTIONS: I have too much work to do everything well; I have enough time to get the job done; I receive enough help and equipment to get the job done; The pace of work increases with automation.

12.2.1.7 Control of Organizational Functions

The questions contained in this section determine the concerns of individuals formerly tasked with supporting the hand operation who will now be tasked with supporting the added complexity of the automated system.

SAMPLE QUESTIONS: In general, relations in my work place between management and employees is good; My supervisor is helpful to me in getting the job done; I am free from the conflicting demands that other people make of me; Automated systems promote the ability of employees to control elements of the organization more effectively.

12.2.2 Data Compilation

Once the questionnaires have been returned, a number of follow-ups need be performed to clarify answers, complete or explain partially complete questionnaires, and address those that have not returned their questionnaires. The follow-up will provide further answers to the resistance against automation. They will be the ones not returning the questionnaires and will provide an explanation for their failure with the common refrain, “I have real work to do. I do not have time for this.”

12.3 Perception of Ubiquitous Automation

Today’s aerospace factories have been subject to all forms of automation that have affected the lives of the people. Automated systems in the office and on the shop floor have primarily come in the form of information technology. The rapid increase in tasks done by automation has shaped the opinions and attitude of the people using it. It is everywhere, and its implementation has either been successful or difficult; and has had either a positive or negative effect on each employee’s life. The attitude of employees toward automating the D/C operation is shaped by their previous experiences with automation. Some individuals have become overwhelmed by the pace of change to machines and perceive them as a threat to their human dignity.

The potential application of another piece of automated technology to complicate their life and take away work can overwhelm some people. The questionnaire can determine the degree that the perception of ubiquitous automation has influenced their attitude toward additional change. However, it may be apparent that the resistance to further change cannot be overcome by training, education, orientation, and coaching. Therefore, there may have to be a realignment of personnel and tasks to incorporate those more accepting of automation.

12.4 Safe, Secure, and Ethical Automation

Resistance to automation where it has not been previously applied or unfamiliarity with automated systems creates distrust and misunderstanding. Beyond the questionnaire, training programs and orientations meant to reduce job threat anxiety, safety, security, and the ethical application of automation must be addressed.

12.4.1 Safety

The safe operation of automated machinery is complicated when applied on-assembly in the midst of a variety of hand operations. The automated D/C machines are operating in and around people. Compounding these legitimate concerns is the lack of experience with safe operation of automated systems in factories that have never dealt with automated systems. Rumors abound of runaway robots with resultant maiming, injury, and death; not to mention the destruction of product. They are legitimate concerns that must be addressed, and some of them are valid. Any machine operating in close proximity to humans must have designs that incorporate safety systems that protect the people near or inside the automation field of operation.

It is imperative that these concerns be addressed as part of the initial orientation, interview, and questionnaire pre-assessment evaluation S-CONOPS. The safety concerns of production operations personnel are important enough that they are addressed separately as an independent survey. The development of the safety questionnaire survey involves the factory resident safety engineer and incorporates the orientation of the safety engineer. The safety engineer for the factory where automation is proposed is briefed by the automation team on the state, local, and federal laws regarding integrating automation into the proximity of people in the work cell. The safety engineer is also briefed on the safety systems that are being used to meet the state, local, and federal laws and regulations. Where possible, trips to factories where automation has been installed are arranged for the safety engineer so they can observe the safe operation and safety systems in operation. Where not possible, the safety engineer is shown video examples of safety systems incorporated on active automated D/C machines integrated into the assembly process. They are also shown the compiled safety data from operational machines. The compiled data has and will substantiate the claims of a safer environment in an automation cell for the same hand operation.

The safety engineer is closely involved in every aspect of the development of survey questions for the safety questionnaire and also in evaluating and educating the individuals that will be involved in the automation process. My experience has shown that a level of trust is given with regard to the operation and efficiency of automation from the factory operations personnel to the automation team, who they are unfamiliar with. But safety claims and briefings are suspect unless they come from one of "their own" who is considered independent and trustworthy with regard to the important safety concerns of the employees. The early integration of the factory safety engineer in the pre-assessment quells fears and concerns early in the incubation phase before feasibility analysis is performed.

SAMPLE QUESTIONS: The safety of workers is a high priority with management where I work; There are no significant compromises or shortcuts taken when worker safety is at stake; Where I work, employees and management work together to ensure the safest possible working conditions.

12.4.2 Security

When automation is installed and operating on the production floor, there will be times when the factory goes lights out and the systems are shut down. Security systems are necessary to minimize damage to the machine or product from curiosity, accidental, or malicious actions that operate the machine outside normal operation by authorized personnel.

Factory personnel are briefed on these security systems so that leadership, customer, quality, and other stakeholders can be provided information to reassure them that the machine cannot be operated outside the ring of authority. The ring of authority is composed of those individuals who are authorized to operate the machine at the times specified for its operation. Included in the security system is a time lock that shuts the machine down automatically should the authorized operator leave for an extended period of time and neglect to power down the machine and return it to the home position.

12.4.3 Ethical Application of Automation

Ethics for automated machinery includes the application of vacuums and filter systems that reduce or eliminate contaminants produced through D/C. Environmental ethics also includes the collection of coolant residue on the tooling, machine, and product as well as the collection of mist to reduce airborne contaminants. The collection of airborne contaminants becomes more important with automation because they use mist coolant that can migrate into the atmosphere of the factory. The potential airborne contaminants contain both the coolant and the dust from the materials that have been drilled. Therefore, just as the safety engineer is a valuable partner in the initial drive to pacify the workforce and mitigate safety concern, the environmental engineer is also brought onboard early in the process. They become acquainted with pre-implementation tests that certify the machine's ability to remove contaminants from the atmosphere and familiarize them with disposal companies and disposal methods for the collected contaminants that meet state, local, and federal guidelines.

The ethical selection of operators is also addressed. The human resources department is involved early to assuage the fears of favoritism for selection of what has become a higher paying automation operator classification requiring less physical effort than the other hand tasks in the work center. Human resources helps the cost center understand the effect of automation to job descriptions within the cost center and assure them of objective selection for operators should automation be installed.

12.5 Management of Automated Systems

Automation is a significant risk for managers and senior leadership. The hand process for the fabrication and assembly of airplanes is well understood. Change to the process comes at a risk; the risk of the unknown. I have experienced leadership that supported automating a D/C process almost to the point of turning the machine on and drilling the first automated unit and then canceling the project. Airplane subassemblies are expensive, and loss of a subassembly has cost the career of a few senior leaders in a company.

Therefore, it is imperative to communicate early, communicate often, and communicate thoroughly as a weekly rhythm along the path to the first hole. Communication and orientation begins long before the feasibility study. Any gap or lapse in the communication stream begins to percolate the fears resident in risk to the reputation and career of the sponsoring leadership. One of the social impacts to leadership is the risk stress associated with the disruptive technology of on-assembly automated D/C.

12.6 Organizations

The social impact to organizations of on-assembly automated D/C systems affects every aspect of the factory, from the data collected for EVMS that had formerly relied on mechanic input for accuracy to purchasing and standard tooling. It is cost prohibitive to engage all of the people that populate these organizations in the pre-assessment orientation. However, targeted individuals and organizations should be selected as advocates to "spread the word" and defeat the emergence of unsubstantiated fears before the evaluation process to determine the viability of automation can begin.

Where labor unions exist, they should be employed and engaged to become partners in the process of advocating for at least the feasibility study and cost benefits analysis. Labor unions have been a major player in the promotion of automated systems to maintain the United State's and Europe's competitiveness in a time when low-cost labor is beginning to make inroads into the airframe fabrication and assembly.

Labor unions understand that fewer but higher-paying automation jobs keep work at home for union members when the absence of automation may drive the hand-performed tasks offshore. The additional but different job descriptions for design, build, operation, and support of automated systems provide a product that allows the United States and Europe to compete.

They are a critical ally for advocating the positive benefits of automation to the workforce and easing the job of the automation application engineers when they arrive to do the assessment.

The early engagement and communication by the automation assessment team before they begin the assessment phase of automated D/C implementation determination has another advantage. It creates familiarity with the factory population they will have to engage when they begin assessing the feasibility of applying automation. The team will be familiar with the people necessary to support the effort. And the factory production operations personnel will be familiar with them.

This was one of the mistakes I made when I introduced myself to the mechanic in Florida. He did not know who I was, why I was there, or what the impact of my presence would be to him or his livelihood. I am convinced that had I applied the principals previously described to address the social impact of the technology I was evaluating for installation, I might have been successful.

12.7 Customers

Customers are the premier stakeholders in the manufacturing process. They are the direct benefactors of the improvements in cost and quality derived from the automated D/C process. Before and during every step along the path to product delivery, the customer has to be engaged. The biggest advocate for change to reduce the stress and risk sensitivity of leadership is the support of the customer. They lead the charge to improve the product and, if engaged early in the process, can help ease the pain of delays and mistakes that inevitably occur along the way to automating product manufacture. The customer's early notification of the desire to spend money to improve processes drives the acceptability of automating processes.

12.8 Summary

Evidence suggests that the effect of automated D/C technology on employees will not always be positive. However, if a major goal of automated D/C is to maximize resources to effectively serve their customers and one of those resources is human resources, then an automated system that predominantly impacts negatively on personnel is a waste of resources such as expertise, ideas, skills, and knowledge; a waste of human potential. Therefore, it is important for management not just to **automate** (the operational aspects of their systems) but to work toward **integrating** the system into the working environment to benefit both management and employees. To realize this goal, those involved in implementing automation technology need to incorporate as part of the implementation process the following basic societal impact principles, which are based on participative management techniques.

12.8.1 Communication

Employees must be kept informed of the progress of the implementation process to avoid feelings of alienation and powerlessness over the change process.

12.8.2 Participation

In addition to being informed, employees must be able to participate actively in the implementation process. Personnel need opportunities to discuss the implementation process, ask questions, raise concerns, and provide feedback to those assessing the potential for automation and implementing the technology.

12.8.3 Consultation

The whole concept of implementing an automated system revolves around changing the ways in which tasks have been done. To maintain employee quality of working life and job satisfaction, *it is not sufficient to superimpose one system on another; due attention must be paid to job design* [12.4]. Leadership, in consultation with employees, must design jobs that contain all or some of the following elements: variety, autonomy, responsibility, feedback and recognition, social contact, discretion and control, achievement, and opportunities to learn and develop. As Austen [12.5] stated,

*The tendency to try and **automate** the person must be resisted. It is all too easy to fit the person to the system, rather than to **humanize** the system—but it is courting trouble! [12.5]*

12.8.4 Training

Besides the initial training employees receive in the use of the system, they also need access to appropriate levels of training as and when required. Failure to provide ongoing training may result in lack of interest, frustration, and inability of employees to realize the full potential of the system to meet their or the users' needs.

12.8.5 Support

The provision of quality and timely support to employees who are having difficulties with accepting the system, allows leadership to feel confident that they can use the system to its full potential. If support is not forthcoming, leadership tends to feel they have little or no control over the system. This can lead to frustration and stress as leadership and employees doubt their ability to cope with the system.

No studies have been done to determine the long-term effects of automation on aerospace personnel. However, *technology cannot determine what choices will be made for what purpose* within the organization, only people can [12.6]. Consequently, when corporations and their leadership implement automated systems to utilize both technological and human resources to their full potential, the negative effects of automation on aerospace assembly operations are minimized. As one worker in Zuboff's *In the Age of the Smart Machine* [12.2] mused,

> *If you don't let people grow and develop and make more decisions, it's a waste of human life—a waste of human potential. If you don't use your knowledge and skill, it's a waste of life. Using the technology to its full potential means using the man to his full potential. [12.2]*

All of the actions described in this chapter come with monetary cost. The cost to overcome resistance without the education and engagement of the affected people, organizations, and environment, which automated D/C relies on for success, is much more expensive.

References

12.1 Mayer, Frédérique, Chair IFAC Technical Committee, TC 9.2, "Social Impact of Automation," Scope members News Activities, http://tc.ifac-control.org/9/2, Accessed May 16, 2012.

12.2 Zuboff, S., *In the age of the smart machine: The future of work and power,* Oxford, Heinemann Professional, 1988, p. 414.

12.3 Center for Disease Control, Workplace Safety and Health Topics 2012, "Quality of Worklife Questionnaire," http://www.cdc.gov/niosh/topics/stress/qwlquest.html, Accessed April 15, 2012.

12.4 Dyer, H., and A. Morris, *Human aspects of library automation*, Brookfield: Gower Publishing Co., 1993, p. 185.

12.5 Austen, G., "What about the bosses?" *Cataloguing Australia*, 13 (3):131–137, September 1987.

12.6 Zuboff, S., "Automate/Informate: the two faces of intelligent technology," *Organizational Dynamics*, 14 (2:5–18).

Chapter 13
Future State of Airframes and Automation Summary

When the only tool you own is a hammer, every problem begins to resemble a nail.

—Abraham Maslow

13.1 Introduction

When the only tools you own are for fastener installation, every problem begins to resemble a hole. The modification of Abraham Maslow's famous quote is appropriate for the manufacture of commercial and military airframes [13.1]. However, the winds of change are blowing and will affect the segments of both types of airframe. In addition to recent developments in material types, the factors of economics, function, and customer requirements are driving the near, medium, and long-term time-horizons for change. For the past 50 years, commercial and military airframes mutually have benefited from development and advances in manufacturing technology. Most of the airframe manufacturers produce both. Therefore, any advantageous development in commercial manufacturing migrates to the military side of production. Development of new manufacturing methods and technology for military airframes migrates to commercial airframes. The ties that bind these two diverse airframe types will continue to a large extent. However, recent developments in UAVs for military applications will drive the development of unique materials, technologies, and processes for their manufacture. Absent a pilot, the airframe and its manufacturing requirements for inhabited and uninhabited air vehicles will diverge into separate technologies and processes. Economics will drive change in how, where, and when airframes are produced; and how many of them are made. The airframe's manufacturing affordability will be a key player in the future of airframe assembly methods.

Holes, hole preparation, and fastener installation are by far the largest contributors to cost, schedule, and quality for an airframe's assembly. There are three primary ways to reduce the cost of holes, hole preparation, and fastener installation. They are:

1. Apply automated systems to D/C faster to reduce time.
2. Eliminate the need for holes.
3. Move assembly operations to low-cost labor markets to reduce cost.

This chapter will explain the effects of economics on airframe assembly, define the key economic contributors that drive change, and identify affordability issues that drive the manufacturing exodus to low-cost labor markets. Technology changes in material and manufacturing requirements for low-cost solutions to replace high-cost drivers such as the autoclave will also be explained.

13.2 Economics

One common thread that runs through commercial and military airframes is economics.

Economics will continue to drive the future state of airframes and automation. Every product reaches a price point where marginal cost exceeds marginal revenue. There is a point beyond which rising cost does not justify the value, genuine or perceived, of the product the customer is pursuing [13.2].

Modern airframes have reached the equilibrium point where the cost of building them has exceeded the ability of most nations to afford them. The nations that can afford them are only able to buy them in very low numbers produced over very long time horizons. Airplanes and their cost drivers have exceeded the economic equilibrium point. Their price point has exceeded the economic ability of the customer to afford the product. Therefore, the buyers of airplanes have incentivized airplane producers to find alternatives or substitutes. Companies and nations are vigorously pursuing alternatives to either replace the current type of airplane or to reduce the key contributors to the cost of their manufacture.

One cost contributor is the sensitivity to change of a modern airframe assembly line. The acceleration of threat varieties in the military market and the need to service divergent markets in civil aviation have challenged the traditional assembly line's capability. The current aircraft assembly line is an evolutionary product derived from a manufacturing system developed before and during World War Two

(WWII). It is meant to be set up, stabilized, and remain in place with a minimum amount of change during its useful life. Economics, today's threats, and demand for tailored products require rethinking the traditional production lines for airframes to satisfy today's customer. They need to become more flexible and capable of assembling multiple airframes on one assembly line.

Airframes have also become more complex and the assembly of their components increasingly more precise. This has also driven the cost of manufacturing an airframe and stretched the ability of the manufacturers to meet the demands of the design criteria. Embedded sensors and antennae in both commercial and military airframes as well as stealth requirements for military airframes have driven up airframe manufacturing costs.

Modern airplanes have reached a critical economic decision point. Modern airplane producers have begun to investigate and develop ways to provide the needed resource (airplanes) at a price point that can incentivize their customers to continue to buy their products.

Economic incentives will drive alternative sources for the manufacture of airframes using traditional methods. The natural action that occurs when faced with a threat is fight or flight. Airframe manufacturers faced with the threat of declining orders respond the same way. They fight with the arguments that the airplanes are more complex and have better performance, which drives cost. Customers care about performance, but when price is higher than the capability of the customer to pay for the product or can only afford ineffectual numbers, performance becomes mute.

So manufacturers default to flight. Flight takes two forms to reduce cost—either become more efficient or reduce the cost of labor. The flight to efficiency drives manufacturers to dig deep to find ways to produce product faster to run away from rising costs. Lean manufacturing and automated D/C appeared on the scene to accelerate the production process. The response to rising cost was not to change the product but to produce the contributing cost drivers faster, therefore lowering the labor cost.

Recent developments in countries where airframes were not formerly produced have opened manufacturing options for lower labor production. At one time the complexity of manufacturing airframes had risen to the point where only a handful of wealthy industrialized nations had the resources and infrastructure needed for their production.

13.3 Manufacturing Exodus

At the end of WWII, the United States reigned supreme in many industries. The United States' airplane production was the sole source for most nations to acquire their commercial and military needs. The exceptions were the Soviet Union, Eastern Block countries, China, and North Korea. However, even these exceptions were trying to acquire the United States' technology. The superiority of the United States aerospace industry continued for the next two decades into the 1960s [13.2].

I remember watching a news program in the late 1960s. The German Minister of Industry said, "The English make the best engines; the Americans make the best airframes; and the Germans make the best ashtrays."

Today's European aerospace and space industry is challenging the United States in many areas. However, aerospace products and technology remains an economic bright spot for the United States. They also are one of Europe's economic positives. Both the United States and Europe have positive export balances for air and space vehicles and technology [13.3].

However, the seeds of change were planted in the early 1970s, when aerospace companies established Offset and Countertrade departments. The Offset and Countertrade departments are established to broaden the market for sale of United States and European airplanes to countries that do not have the hard currency for their purchase. At first, the operations and transactions of the Offset and Countertrade departments were comical in their simplicity. In the early 1970s, at the Convair facility in San Diego, employees working on the DC-10 fuselage received a memo asking them to buy Polish hams. To make the purchase of Polish hams easier, employees were "asked" to use the payroll deduction method to acquire a minimum of one ham per week. The Polish wanted DC-10s but lacked enough hard currency to buy them. The Poles had a surplus of hams. The Douglas Offset department would find markets for hams in countries (and from employees) that had hard currency. Hams would become dollars, and the Polish government used the dollars to buy a DC-10.

The idea of offset and countertrade caught on [13.4]. In industrialized Western countries, the exchange became more sophisticated. In the early 1970s the Swiss, notorious for their negotiating skills, bought the F/A-18 C/D Hornets because Boeing and Northrop agreed to teach them to build a segment of the airframe. Other competitors were trying to sell their airplanes to the Swiss. Switzerland built the F/A-18 C/D Hornet vertical stabilizers to offset the cost of the airplane. The technology exchange played a major role in winning the contract. Northrop Grumman Corporation (then Northrop Corporation) began transferring a new F/A-18 C/D Hornet vertical stabilizer assembly line complete with planning, blueprints, engineering, training, and anything else it would take to make vertical stabilizers. Today, the Swiss aerospace company Ruag is a producer of many air and space components that had their genesis in the transfer of the manufacturing technology as part of offset and countertrade agreements.

Offset and Countertrade departments have evolved to become integral to the success of aerospace companies today [13.5]. Offset and countertrade are highly sophisticated but retain key elements for expanding markets for aerospace products into countries without enough hard currency for their purchase.

Offset the cost of the sale to Malaysia for F/A-18 C/D Hornet airplanes by finding buyers with hard currency for palm oil. Entice the British to buy F-35s by giving them the forward section to build. Teach India or China to build Boeing 777 or 787 subassemblies for shipment to assembly centers in the United States on condition that they buy Boeing airplanes. Emerging economic powerhouses such as India and China are fast learners. While in the past a few manufacturers and even fewer nations could produce airframes or components, today the capability is spreading across the globe to include low-wage airframe producers for a highly labor intensive product.

Two recent examples are the Harbin Hafei Airbus Manufacturing Center in China [13.6] that will fabricate composite parts and assemblies for the A350 XWB, Airbus' latest aircraft program. In India, Boeing not only builds composite parts for their 787 but also has established a relationship with the Indian Institute of Science for research in aerospace material, structures, and manufacturing technologies [13.7].

Unless anyone thinks military hardware is exempt for anything but the most advanced military aircraft, assembly lines for components and assemblies for the Lockheed Martin F35 are setting up in countries such as Turkey and Australia [13.8].

In the 1980s, the venerable Teledyne Ryan AQM/BQM/MQM-34 Firebee UAV airframe assembly moved to Mexico. Figure 13.1 shows the Firebee UAV. The Firebee UAV is one of the most successful and versatile unmanned (drone) aircraft. The latest versions are still in widespread service more than 50 years after the first flight tests.

The flight to low-priced labor markets has begun for the airframe. New composite materials and processes are in development that will replace the autoclave and accelerate the exodus of airframe assembly from high-priced assembly centers in the United States and Europe. An autoclave is a major barrier for entry into the airframe business. Airplane and space components and parts require an autoclave to perform the process to cure and compress composite parts. Prepreg-grade composites require an autoclave for several reasons. Compression and consolidation are two. Another is that the autoclave acts as a pressure cooker, preventing the boiling of moisture resident in the composite by raising the boiling point above the cure temperature of the composite material. Autoclaves are very large pieces of equipment. They are expensive to acquire and operate. Nitrogen, natural gas, vacuum lines, and thermocouples add complexity and cost to the process. Figure 13.2 shows the size of an autoclave and its intrusion into the manufacturing factory layout.

Fig. 13.1 The Firebee UAV.

Fig. 13.2 A typical autoclave.

Autoclaves are immovable objects that defy reconfiguration or movement to a new facility. Therefore, they are a barrier to entry for potential low-cost-labor airframe producers other than wealthier nations such as India or China.

13.4 Exit the Autoclave

The economic impact of eliminating the autoclave is significant. Composite parts that do not require an autoclave are produced at widely disbursed low-cost facilities rather than in large groups at a centralized location. Grouping of autoclave cured parts together is necessary because of the cost of operating the autoclave. The autoclave costs the same to run with one part as with many parts.[1] Therefore, autoclaves drive fabrication scheduling to realize the greatest number of parts per run. Elimination of the autoclave removes this restriction. Parts are produced without the large intrusive autoclave on the factory floor.

Elimination of the autoclave also enables the design of larger parts to form unified structure. The size of the autoclave limits the size of parts. The unification of parts is not unknown, and current automobile design incorporates unified structure to reduce cost and assembly steps, and increase quality. Airframes would benefit from the same design-for-manufacture attributes realized by the automotive industry for the last three and one-half decades. Unified structure also eliminates fasteners that require holes. Holes are a major cost and quality driver in airframe assembly. Examples of unified structure exist that reduce fasteners count and, therefore, hole quantity. The Boeing H-47 Chinook helicopter team unified one composite part on the under nose section, reducing fastener count from 4000 to 400.

Elimination of the autoclave will also remove a major barrier to entry for additional low-cost labor producers to enter the airframe part manufacturing arena. Removal of the autoclave cure process also reduces airframe part fabrication complexity and further enables low-cost labor entry into the part fabrication market. Lower-cost unified structure produced in lower-cost labor markets incentivizes the acceleration of airframe designers and assemblers to acquire parts outside the United States and European markets. Reduced-cost unified structure produced below major airframe assemblers shop rates of $150 to $200 per hour would provide a significant airframe cost reduction.

Two major thrusts are in progress to eliminate the autoclave. One is Liquid Resin Infusion (LRI), and the other is Out of Autoclave Cure (OAC) of autoclave cured material [13.9].

The liquid resin infusion process involves laying dry fiber into a mold, applying vacuum, and pulling resin through the fibers (infusion). Compared to the autoclave, it is a relatively simple process and has been the composite manufacturing method

1. Purists will argue that a smaller part has a shorter cure cycle than a large part and therefore costs less to run. The delta energy cost is marginally small when compared to the cost of starting, heating, and running the autoclave, either full or empty.

and material of choice for industries such as ship building and wind turbine blades. Over the last two decades, the material properties for LRI have improved to the point that they are used on air and space vehicles. As they continue to improve, it is anticipated that their properties will align close enough with high-grade prepreg to offer a low-cost alternative for most airframe applications [13.10]. In addition to lower labor cost and manufacturing complexity, the added value of LRI is material cost. LRI material cost per pound is a fraction of the cost of aerospace-grade prepreg autoclave cure material.

The other alternative is for aerospace-grade autoclave cure material out of the autoclave. The goal of the process is to retain the autoclave material's strength properties while eliminating the expense of acquisition and operation of the autoclave. One of the autoclave's functions is to prevent moisture contained in the composite material from boiling as the part goes through the compaction, consolidation, and cure cycle. The approach used for curing the autoclave cured material out of the autoclave is to place the part in a vacuum bag on the mold. Vacuum is then drawn to compact and consolidate the material while raising the temperature slightly above water's boiling point of 100°C (212°F), but below the cure temperature of the composite material. Vents bleed off the moisture in the material, and the temperature rises to cure and solidify the part. This process requires an oven. An oven is less expensive and more portable than an autoclave.

Both of these processes produced parts for the Max Launch Abort System (MLAS) that the National Aeronautics and Space Administration (NASA) launched from their Wallops Island facility in Virginia. The fabrication process for the MLAS fins was OAC of autoclave material. The body parts composed of the boost skirt, coast skirt, and forward fairings were built using liquid resin infusion. Figure 13.3 shows the successful launch of the MLAS from Wallops Island, Virginia, on July 9, 2009.

Fig. 13.3 MLAS Launch.

The combination of these two types of processes that provided components for the MLAS launch vehicle was accomplished in low-cost environments without benefit of high-cost facility investments. The production techniques, processes, tooling, and quality control are transferable to low-cost labor environments.

The cost of airframes is driving their price point beyond the customer's ability to pay.

Advances in technology may facilitate the demise of one remaining vestige of a robust United States and European manufacturing industry. There are examples in which the survival of a manufacturing industry was salvaged from the scrap pile at the sunset of their operation. There are examples of factories left vacant by failure to heed examples of success.

13.5 Automation or Labor

A colleague at Oxford University arranged a private tour of the Mini automobile manufacturing facility in Oxford, UK. The former plant manager explained that the facility had two options, automate or close. They replaced 7000 workers with 700 robots supported by 1500 technicians. The importance of keeping the factory open and the Mini brand staying at home in the UK, even with the loss of 5500 jobs, was a critical decision. The new jobs were higher paying but still only a fraction of the jobs the factory previously needed to build the Mini automobile by hand.

At a time when jobs are at a premium and economies are struggling, the call often goes out for jobs. The employment of people overshadows the need to provide manufactured goods to a market that needs and wants the product. Aerospace is not any different from many car companies such as Range Rover and Jaguar that have migrated to lower-cost-labor manufacturing countries. The future and survival of both commercial and military airframe manufacturing in the United States and Europe may very well depend on the same decision Mini and the town of Oxford had to make—automate or die. The backlash against automation derives from loss of jobs to machinery. In fact, the refusal to automate has cost many more jobs to low-cost labor markets than automation. If a manufacturing firm's labor is unwilling to compromise and automate processes, and cheaper substitute products exist in the marketplace, buyers will migrate to the lower-cost product. The result will be that the firm will operate at a loss until its position becomes untenable. As a result, the firm will close, losing all the jobs instead of those replaced by automation. A successful alliance between labor and automation is the Bavarian Motor Works (BMW) manufacturing facility in Dingolfing, Germany [13.11]. It is a highly automated line in the middle of one of Europe's highest paid labor regions. BMW employs 18,000 people and 500 apprentices at this highly automated and flexible assembly line. The line combines three models of automobile in almost infinite configurations on one production line that produces 1500 cars a day. When other automakers' factories in Europe were fleeing, selling, or closing, BMW automated, survived, and thrives.

Besides enabling high-priced-labor countries to compete against low-cost-labor automation has transformed the definition of quality. The modern automobile is a high-quality product. Its superior quality to previous generations of automobiles derives from the capabilities of automation. Artisans and craftspeople used to produce automobiles that carried the variability of the hand fabrication method into the assembly process. As cars produced by automation swept from Japanese automobile manufacturers into U.S. markets, consumers became used to higher-quality products. This forced the United States auto manufacturers to change through market force incentives.

An example of automation's impact on the airframe quality was the automated drilling of the T-38 wing at the beginning of 2000. Northrop Grumman Corporation (NGC) was one bidder to re-wing more than 400 of the T-38s. NGC was the designer of the aircraft and considered the T-38 a legacy airplane. NGC wanted desperately to win the contract to re-wing the trainers it had designed, built, and first flew in 1969. Many low-cost labor shops are available that could produce the T-38 wing. NGC knew the price point that could win the contract but could not produce the wing at the needed price. Automation offered the answer to the dilemma, and the development of an automated T-38 wing drilling system began. Once the T-38 D/C machine was operational, test runs were made on scrap wings to confirm process times to validate that NGC could meet the cost and price bogies. The T-38 wing is an automation dream. The lower skin is open, flat, and has thousands of holes equally spaced in long rows. The skins are aluminum covering aluminum substructure, and therefore would be suitable for high speeds and feeds for rapid, one-shot hole drilling. The upper skin is the same, but has a slight contour. One difficulty was developing drill routines for the automation from archived microfiche images on microfilm. Another was overcoming history and urban legend.

Over time the hand manufacturing method had introduced stresses into the wing that warped the wing. The United States Air Force had to trim the wings before installation on the T-38 to compensate for warp. Over time, the shop legend grew that the warp is intentional for some unknown engineering reason.

Automation produced the holes and countersinks in a scrap wing. The result proved the viability and competitive cost benefits for using automation to produce the T-38 wing. On January 29, 2001, NGC announced it had initiated production of 55 replacement wings for the U.S. Air Force's T-38 Talon supersonic trainer. Deliveries began in October 2001.

Upon delivery of the first wing produced with automation, the U.S. Air Force set up the aircraft to trim out the wing after installation as part of the age old acceptance and preflight procedure. The first wing and every wing delivered afterward no longer needed trimming. Automated D/C had eliminated the warp created by the hand-built process. This was an added cost and schedule benefit to the U.S. Air Force.

Subsequent analysis revealed that the variability of the holes drilled by hand caused stresses during fastener installation that in turn warped each wing differently. Wing warp is built-in by the hand manufacturing process.

The introduction of automation produced uniformly consistent, high-quality holes that reduced the stresses during the fastener installation process. As a result, production and assembly quality improved and resulted in a higher-quality wing to be delivered, installed, and ready for in-flight use. An additional benefit derives from improved fit-up and installation of the wing. The wing no longer requires the adjustments of previous wings to install the attach bolts.

Automated D/C reappeared on the airframe assembly scene in the 1990s. Earlier technology was not mature enough to meet the manufacturing standards required to assemble the airframe. Modern controllers for automated machinery and more precise machinery reincentivized the application of the automation to the assembly floor. The incentive to reintroduce automation is the desire to drill holes faster. Holes are the gateway operation to fasten the airframe together. Therefore, drilling them by machine at an increased rate is a primary option to reduce assembly cost.

Aerospace is on its third generation (GEN3) of automated drilling machines integrated onto or into airframe assembly lines around the world. Ready for use are fourth generation (GEN4) drilling machines that incorporate flexibility and lower-cost solutions than the previous generations of D/C machines. They are an evolution of low-cost, articulating arm robots that began as pick-and-place robots in the automotive industry.

The improvement to positional repeatability has elevated them to the capability to do precision operations such as D/C. The lower purchase and operation cost, and smaller size when combined with flexible applications offers further incentive to automate.

In addition to unified structure, the costs associated with fastening components together began to incentivize examination of other alternatives to fasteners. The cost of fasteners, the process associated with their installation, and their weight began a hunt for alternatives.

Two alternatives that have shown results and are migrating into today's modern airframes and eliminating the need for fasteners are Click Bonds and Z-pins.

13.6 Fastener Alternatives

Unified structure is one way to eliminate fasteners. It is the most visible and, therefore, receives the most attention. Internal fasteners are eliminated by two processes that continue to grow and migrate onto the airframe to reduce fastener count.

The primary disadvantage for airframe manufacturers is the assembly time required to install riveted nut-plates.[2] Nut-plate installation includes the precision D/C of two rivet holes per fastener to affix the nut-plate to the airframe for a variety of functions.[2] These two holes are in addition to the primary attach hole for the mating bolt.

2. Removable panels are bolted onto substructure using nut-plates.

Areas that require additional holes to affix internal components are those that attach clips and brackets. Clips and brackets hard mount to the substructure with rivets and fasteners to hold internal components such as wire harnesses.[3]

A product called Click Bonds is replacing nut-plates, clips, and brackets. More than a component, they represent a process included in the purchased component. Adhesive bonded nut-plates, clips, and brackets eliminate the rivet holes by replacing the rivets with adhesive as the method of attachment. These techniques eliminate the need for many of the primary structural fasteners, and the material-damaging fastener holes. These structures are lighter and perform better than their riveted or spot-welded counterparts, and the weight savings results in cost savings for aerospace applications. Click Bonds are used on the Boeing C-17 and F/A-18 E/F [13.12]. The Click Bond system in the C-17 has eliminated the need for an estimated 2,784,000 holes thus far.[4]

Z-pins are composite fiber rods that act like thousands of nails to connect two composite pieces and create an extremely strong bond after curing. Labor and weight savings have been substantial. Z-pins have eliminated 4800 fasteners and 48 pounds per ship set on the F/A-18 E/F Super Hornet. The Z-pin process also eliminates the nylon channels formerly needed to cover the fuel tank fasteners [13.13].

When combined with unified structure, Click Bond and Z-pins have the potential to eliminate substantial numbers of fasteners and reduce the cost and complexity of airframe assembly.

13.7 Conclusion

Economics will continue to drive the future of airframe assembly. Performance is essential to meet the demands for fuel efficiency and speed. However, the demand for performance is increasingly offset by the ability of the buyer to afford the product. The airplane price "pushback" has incentivized manufacturers of modern airframes to find low-cost alternatives to producing airframes.

Several solutions can and do reduce the cost to produce an airframe. Each has its benefits and all are migrating into the fabrication and assembly of airframes. The trend toward low-cost solutions will continue into the future. Low-cost labor will drive the migration of part production and eventually entire airframe production to low-cost-labor producers. The first business type to migrate to foreign production and eventual ownership will be private, business, and corporate airplane and helicopter manufacturing so intensely vilified by the United States and European Governments. Government policy and tax burdens toward small business airplanes and helicopters in the United States and Europe will push production and eventual sale of a once vibrant business to low-cost-labor countries [13.14]. Production

3. Wire harnesses and other internal components, such as tubing, are called "stuffing."
4. Based on the 232 C-17s produced as of June 2011.

migration will occur to offset the cost to produce and operate the business airplanes and helicopters in the United States and Europe. The production and sale of business jets accounts for more than 1,000,000 high-paying [13.15] manufacturing jobs and supports more than 200,000 jobs of other types, including engineering, support, and control. Some of the jobs will remain, but the paradigm will change to the lease business model or the nation of registry business model, further eroding the United States and European airframe manufacturing base [13.16]. The lease model is already migrating into the United States and Europe for business and corporate airplane and helicopter use. Individuals and companies are beginning to lease their small planes for personal and business use from other nations, such as Brazil. Brazilian companies register, own, and maintain the plane and crew out of Brazil but operate the plane in the United States or Europe. Brazil has a thriving aerospace industry that offers not just an alternative to expensive United States ownership, but also an alternative to expensive United States and European manufactured airframes and airplanes [13.17].

Large commercial airframes and eventually entire airplane types will follow the small business and private plane migration to other countries offering low-cost labor. Accelerating the migration will be the maturation of simplified unified structure cured out of the autoclave. Large unified structure reduces the complexity of the assembly process by eliminating holes and fasteners. OAC reduces the complexity of part fabrication. Combined, these two fabrication and assembly elements require less skilled labor to produce an airframe of acceptable quality.

Composite materials have had a major impact on the alternatives to assembly. Bonding and fastening alternatives, such as Click Bonds and Z-pins, reduce the complexity of the process and therefore reduce the skill required by the mechanic to assemble the airframe. When added together, OAC, unified structure, nontraditional fasteners (Z-pins), and glued piece parts (Click Bonds) will simplify the entire airframe fabrication and assembly process. The major effects will be a lower-cost skill set, resident in low-cost labor nations capable of producing acceptable quality airframes and eventually the entire airplane. The result will be a lower-cost product available to a wider market that will reinvigorate the wide distribution and use of airplanes.

Military airframes and airplanes will follow. The process of migration to lower-cost markets for the component fabrication and assembly of United States and European military airplanes has already begun. Two recent contract awards, the new U.S. Air Force tanker and the new U.S. Navy Poseidon aircraft are derivatives of commercial aircraft. The United States tanker contract was won by Boeing, beating out the EADS after much flag waving. The airplane has Boeing 767 engines and airframe. Thirty percent of the airplane parts and components originate outside the United States [13.18]. The new U.S. Navy P8 Poseidon uses the Boeing 737 airframe that has many of its components made in low-cost labor nations. The U.S. Air Force E-4 and E-8 types use the 707 airframe designed and first built in 1958.

The performance of many modern military airplanes is becoming more a product of electronics and added materials such as LO coatings than the design of the airframe. Much of the airplane can be produced in the "white world"[5] and finished in a secure "black world" environment without compromise to the classified mission or capability of the airplane or helicopter. The incentive to European and United States defense needs is significant. Design and finish in Europe or the United States and build in low-cost labor nations reduces defense expenditures, increases weapon system affordability, and maintains force quantities effective for defense.

Increasing numbers of UAVs will influence the future of military airframe production. UAVs are migrating into military use and will migrate eventually to commercial use in both rotary and fixed wing types. They will both simplify and complicate the airframe assembly process.

Sitting in a van near a runway in the high desert near Palmdale, California in 2010, I watched an F-16 approach the end of the runway at near stall speed. Below the F-16, as it flew over the end of the runway, was a Global Hawk UAV. For the first time, it would be allowed to fly over previously unrestricted public airspace. As the chase plane approached, the Global Hawk released its brakes and began to accelerate down the runway, left the ground, and began to fly. An hour earlier, I had been in a hanger where the first Navy Unmanned Combat Air System (NUCAS) was under construction. Earlier in the year, I had been in Mississippi where the MQ-8 Fire Scout is assembled. What these different types of vehicles have in common is not just that they are uninhabited, but that they fly without preprograms or human intervention. They are autonomous thinkers and actors. In England at a research facility dedicated to developing autonomous flight, I had witnessed the evolution of unitary flight to multiple vehicles flying and making decisions based on the birds' response to threats. Multiple aerial vehicles were acting in unison in a collaborative effort to accomplish a goal and safely return home.

All of these developments are influencing the airframe and its future assembly process. Without a human in the cockpit, the airplane manufacturing complexity is eliminated for life support and facilitating the pilot's activities. Without the needed life support systems, the airframe can perform activities such as extreme maneuvers that complicate the airframe fabrication and assembly.

Airplanes and airframes will be around for a long time. The B-52 continues to fly and perform. The T-38 will fly through 2040. The state of aerodynamics and airframes that evolved as materials and manufacturing matured together are well understood. Airframes for commercial and military use have matured to provide increasing levels of performance until they have become too expensive or marginally affordable for the buyer. The assembly of airframes accounts for the largest contributor to the cost of the airplane. The drive to lower cost and entice buyers to purchase airplanes is driving airplane producers to seek solutions for

5. "White World" and "Black World" are terms used in the defense industry to refer to open unclassified projects and closed secret projects.

production of future airplanes. Compounding the efforts of airplane producers in Europe and the United States are high labor rates and unfriendly government policies that prohibit innovative solutions and provide disincentives to buyers. Therefore, airframe manufacturers are following the automotive model of the 70s—unify structure to reduce assembly steps, automate to improve quality, and reduce labor cost or flee to low-cost-labor nations for production.

Government policy and resistance automation by labor may leave only one option, flight to foreign lower-cost-labor nations for production. Even flight to lower-cost labor within the United States has met resistance. The International Association of Machinists and Aerospace Workers union District 751, bargaining unit, backed by the national union, on March 29, 2010 filed a complaint with the National Labor Relations Board [13.19]. The suit charged that Boeing illegally put its second 787 Dreamliner assembly line in South Carolina in retaliation for strikes in Washington and should be required to build the line in Washington, according to a potentially groundbreaking National Labor Relations Board complaint.

Given the backlash against in-country low-cost labor production and automation by the government and labor, the future of the United States and European airframe manufacturing may lie in their design and use while moving their production offshore.

European and United States Government and labor constraints, when combined with simplified assembly processes such as unified structure, Out of Autoclave Cure, Click Bonds, and Z-pin technologies, will enable low-cost-labor countries to compete and win airframe assembly contracts.

References

13.1 Maslow, Abraham, BrainyQuote.com, Xplore Inc, 2012, http://www.brainyquote.com/quotes/quotes/a/abrahammas126079.html, Accessed March 21, 2012.

13.2 Landsburg, Steven E., *The Armchair Economist, Economics and Everyday Life*, Free Press: New York, New York, 1993, pp. 10–30.

13.3 U.S. Census Bureau, *The 2012 Statistical Abstract, The National Data Book*, 1309—United States Total and Aerospace Foreign Trade, Table: 1309, 2012.

13.4 Brauer, J. and Dunne, P, *Arms Trade and Economic Development, Theory, policy and cases in arms offsets*, Routledge, London, 2004.

13.5 U.S. Department of Commerce, *Offsets in Defense Trade, Fifteenth Report to Congress*, U.S. Department of Commerce, Washington, DC, 2011.

13.6 China Dailey, "Harbin Hafei Airbus opens new production facilities in China," chinadaily.com.cn, 2011.

13.7 Fetters-Walp, Reic, "Building a strong relationship with India is Crucial to Boeing's growth in this dynamic country," *Boeing Frontiers*, 2010, pp. 22–25.

13.8 Reuters, "UPDATE 1-Turkey keeps plan to buy 100 F-35 fighter jets," http://www.reuters.com/article/2012/02/23/turkey-f35-lockheed-idUSL5E8DN2IM20120223, Accessed March 22, 2012.

13.9 "Autoclave quality outside the autoclave?" *Composites World*, http://www.compositesworld.com/articles/autoclave-quality-outside-the-autoclave, 2006, Accessed March 22, 2012.

13.10 Kleineberg, M., U. Wenner, and M. Hanke, "Cost Effective CFRP-Fuselage Manufacturing with Liquid Resin Infusion (LRI) Technologies," DLR Institute of Structural Mechanics, http://www.dlr.de/fa/Portaldata/17/Resources/dokumente/institut/2002/2002_10.pdf, 2002, Accessed March 2012.

13.11 BMW Werk Dingolfing, http://www.bmw-werk-dingolfing.de/dingolfing/htdocs/lowband/com/de/index.html, Accessed March 2012.

13.12 "C-17 plane wired for big savings with composite brackets," *REINFORCEDplastics*, http://www.clickbond.com/files/Documents/c17_application_article.pdf, 1999, Accessed March 2012.

13.13 *Composites Science and Technology*, http://www.sciencedirect.com/science/article/pii/S0266353804001411, 2004, Accessed March 2012.

13.14 *National Journal*, http://www.nationaljournal.com/obama-s-taxing-corporate-jet-policy-20110629, Accessed March 2011.

13.15 JETHUB, http://www.jethub.com/Aviation-Industry-Blog/bid/79690/Private-Jets-Create-Jobs-and-Help-Boost-Economy, Accessed March 2012.

13.16 Fernandes, Arnold, "The rise and fall of the tuna industry," http://www.westcoasttunahistory.com/, Accessed March 2012.

13.17 Trimble, Stephen, "Brazil aerospace and defense industry evolves, *Flight International*, http://www.flightglobal.com/news/articles/brazil-aerospace-and-defense-industry-evolves-355747/, 2011, Accessed March 2012.

13.18 "World Airline Census 2011," Flight International, Special Report.

13.19 Smith, Bruce, "Labor suit against Boeing a congressional issue," Associated Press, http://www.chron.com/business/article/Labor-suit-against-Boeing-a-congressional-issue-2081529.php, 2011, Accessed March 2012.

Appendix A
Automated Machine Suppliers for D/C of Airframes

Appendix A is a list of automated D/C ***system*** suppliers. It does not include component suppliers to the machine systems such as end effectors, controllers, machine motion components, or other detail elements that compose a machine system. Nor does it include machine suppliers who do not incorporate, as part of their business model, a strategic partnership with their customer to become an integrator into the customer's production line.

1. M Torres
   M. TORRES USA
   2750 South Harbor Blvd., Suite "C"
   Santa Ana, CA 92704

   M. TORRES DISEÑOS INDUSTRIALES, S.A.
   Ctra. Pamplona-Huesca, Km 9
   31119 Torres de Elorz
   Navarra, Spain
2. Advanced Integration Technology
   2805 East Plano Parkway
   Plano, TX 75074
3. Comau Aerospace
   21000 Telegraph Road
   Southfield, MI 48033
4. Electroimpact
   4413 Chennault Beach Road
   Mukilteo, WA 98275

5. Broetje-Automation GmbH
   Stahlstr 1-5
   26215 Wiefelstede, Germany
6. Exechon AB
   Orrvägen 26
   192 55, Sollentuna, Sweden
7. Alema Automation
   Espace Technologique Jean Bertin, Z. I.
   Avenue du 1er mai, F-40220
   Tarnos, France
8. Ingersoll Machine Tools, Inc.
   707 Fulton Avenue
   Rockford, IL 61103
9. KUKA Systems GmbH
   Bluecherstrasse 144 | 86165
   Augsburg, Germany

Index

Access to structure, 66–67, 79, 93–95
Acoustical analysis, 195
Aerospace Automation Consortium, 113
Aerospace Manufacturing and Automated Fastening (AMAF) Conference, 113
Affordable Design and Manufacturing (ADAM), 39
Affordable Systems Optimization Program (ASOP), 39
Air Force Research Laboratory (AFRL), 30
Air supply, 25, 26–27
Airbus crashes, 53
Airframes
  and automation, future of, 14–16, 219–233
  historical perspective, 3, 17–35
    aluminum, 19–20
    assembly tooling, 23–24
    composites, 28–30
    hand tools and drills, 25–28
    on-assembly drilling, 32–35
    rivets, 20–22
    titanium, 24–25
    tooling and composites, 30–32
manufacturing cost drivers, 3–4, 37–53
ALCOA, 19
Aloha Airlines Flight 243, 21–22
Aluminum, 18–20
Aluminum tape ink dot test, 143–146
  skin off, 145–146
  skin-on, 146
Ares I-X, 29
Assembly efficiency, 44–45
Assembly line, airframe, 55–56
Assembly tooling, 23–24
Autoclaves, 32
  elimination of, 223, 225–227
Automation, 4, 55–56
  and airframes, future of, 14–16, 219–233
    automation vs. hand labor, 227–229
    economics, 220–221
    eliminating the autoclave, 225–227
    fastener alternatives, 229–230
    manufacturing exodus, 221–225
  choosing the right automation
    affordability, 110–112
    complexity, 112
    needs assessment, 109
    references, 112–113
    requirements document, 107–108, 113–115
commercial manufacturers' disincentives for
  global geopolitical, 68
  labor, 68–69
  local geopolitical, 67–68
commercial manufacturers' incentives for, 69–70
considerations before automating, 5–6, 91–106
  cost benefits analysis, 100–105
  feasibility analysis, 93–100
  final decision criteria, 105
ethical application of, 214–215
vs. hand labor, 227–229
installation steps, 7–8, 119–137
  factory acceptance, 133–136
  Manufacturing Readiness Level (MRL) consideration, 132–133
  mapping the current process, 120–125
  mapping the process incorporating automation, 125
  performance measurement tools, 133
  platform specification, 128–129
  request for proposal, 129–132
  requirements document, 126–128
  testing, 136
machine suppliers, 235–236
maintenance of, 11, 187–201
management of systems, 215
vs. mechanization, 3
military's disincentives for
  controlled profit, 62
  jobs, 61–62
  lack of competition, 63
  parceled application, 59–60
  politics, 61–62
  production tooling, 57–59
  restricted sales, 62–63
  return on investment, 61
  specific use, 60
  time horizon, 60
military producers' incentives for
  access, 66–67
  affordability, 64
  complexity, 65
  enlightenment, 64

military producers' incentives for (*cont.*)
marketing, 66
multiyear buys, 65
quality, 65
pacifying the human capability-machine experience, 204–212
safety, 213–214
security, 214
Social Concept of Operations (S-CONOPS), 206–212
social impact of, 11–14, 203–218
time study analyses, 101–102
training of personnel, 9–11, 169–185
Transition to Production (TTP), 8–9, 153–167
types of, 5, 73–90
action, 81–86
calibration, 88–89
controller, 86–88
coolant, 85
dust collection, 84
end effector, 81–83
human machine interface, 86–87
inspection, 85–86
integration level I, 81
integration level II, 86
integration level III, 87–88
machine types, 76–80
motion, 74–81
size determination, 75
sustainability, 89
tool changer, 85
tools and parts, 80–81
vision system, 85
volumetric compensation algorithm, 87
ubiquitous, perception of, 212–213
Autonomous maintenance, 197–198

Basic process control system (BPCS) test
error, 135
failure, 135
fault, 134
Bavarian Motor Works (BMW), 227
Bayer, Karl Joseph, 19
Beechcraft Starship, 29
Black metal, 15, 28
Boeing 787, 69, 233
Bonded structure, 14–15
Brazil, 231
Breakdown maintenance, 198

Calibration 11, 88–89
Call-to-site (CTS), 97
Cartesian coordinate system, 178
CATIA (Computer-Aided Three-Dimensional Interactive Application), 35
Certification, 198
Chang, Frank, 34, 78
Chip removal, 31
Cincinnati Lamb, 111
Click Bonds, 230, 231
Cold working, 6
Collision avoidance test, 141–143
Commercial manufacturers
disincentives to automate, 67–69
incentives to automate, 69–70
Communication, 217
Compensation algorithm, volumetric, 87
Composite Crew Module (CCM), 30
Composites, 28–30
dissimilarity of materials, 31
thickness variability, 30–31
and tooling, 30–32
untooled rough surface of, 31
Compression deflection test, 146–147
Computerized Maintenance Management System (CMMS), 193
Condition-based maintenance, 190–198
Consultation, 217
Control of organizational functions, 212
Coolant, 85
Cost benefits analysis, 100–105
ancillary costs, 102
automated drill time study, 101–102
cost collection and conversion, 103–104
deliverables, 105
and final hole count, 161–162
hand process time study, 101
machine cost, 102
return on investment, 104–105
Cost centers, 47, 93
manager training, 183–184
Cost to finance (CFIN), 41
Costs
of automation
ancillary, 102
converting time to money, 103–104
machine, 102, 110–112
defining, 41–42
direct, 37–38, 42
estimating, 131
of hand labor, 37–38, 44–51
identifying manufacturing-related, 43–51
indirect, 38, 42

life cycle, 41
motivation to reduce, 38–40
of quality, 38, 48–50
support, 51
of workforce injury, 38, 50
see also Manufacturing cost drivers
Countersink cages, 23
Countersink depth, 21
test of, 147–148
Countertrade and Offset departments, 222–223
Critical Design Review (CDR), 132–133
Cured ply variability (CPV), 30
Currency exchange rates, 132
Current operations
feasibility study, 97–98
mapping, 120–125
Customers, social impact of automation on, 216

Defects, 48–49
Defense Manufacturing Conference (DMC), 113
Defense Manufacturing Council (DMC), 39
Deflection analysis, 96–97
Demonstration testing, 149–150
Design engineers, 177–180
Deville, Henri Etienne Sainte-Claire, 19
DH 98 Mosquito, 15
Digital holographic imagery, 193–194
Direct costs, 37–38, 42
Disruptive technology, 2
Disruptive Technology Office (DTO), 16*n*
Drill bonnets, 23, 24
Drill cups, 23
Drill routine conformance test, 147
Drill speed, 26–27, 31
Drill starts, 27, 49
Drill stops, 27–28
Drill templates, 23, 24
Drilling, on-assembly, 32–35
Drills, 25–28
pneumatic, 25, 26
DuPont, 24
Dural, 20
Duralumin, 19–20
Dust collection, 84, 150

Earned Value Management System (EVMS), 133
Earthquakes, 189, 190
Economics, 220–221
Edge Distance (ED), 13, 16*n*
Effectivity, 46–47
Efficiency, 44–45
Electronic Gantry Applied Drilling System (EGADS), 33
End effector, 81–83
capability, 83
power, 82
pressure foot, 83
size, 82
tool holder, 83
Engineers
design, 177–180
manufacturing, 180–182
Engineering and Manufacturing Alliance Project Teaming Agreement, 131
Equilibrium model, 39
Error, 135
Ethical application of automation, 214–215
Export control laws, 113

F-35, 66, 70, 110111
F/A-18, 1, 26, 33, 46, 222–223
Fabrication efficiency, 44
Factory Acceptance Test (FAT), 133–136
basic process control system (BPCS), 134–135
error, 135
failure, 135
fault, 134
hazard, 135
instruments, 135
process risk, 135
safety instrument function (SIF), 136
Factory security, 115
Failure, 135
Fasteners
alternatives to, 229–230
failure of, 49
historical perspective, 17
Fault, 134
Feasibility analysis, 46, 93–100
access to structure, 93–95
current operations, 97–98
deliverables, 100
environment considerations, 98–99
factory infrastructure, 97
flooring, 95–96
labor, 99
risk assessment, 99
tooling integration, 96–97, 98
Feasibility matrix, 95

Firebee Uninhabited Aerial Vehicle (UAV), 223
Flooring, automation considerations, 95–96
Flow charting, 121
Flush head rivets, 21

Gantry Applied Drilling Systems (GADS), 78
Generally Acceptable Accounting Principles (GAAP), 112
Geopolitical considerations in automating, 67–68
Glass Fiber–Reinforced Polymer (GFRP), 28–29
Green structure, 5
Gregor, William, 24*n*

Hall, Charles Martin, 19
Hand labor, 177
  vs. automation, 227–229
  cost of, 37–38, 44–51
  time study of, 101
Hand tools and drills, 23, 25–28
Harbin Hafei Airbus Manufacturing Center, China, 223
Harm, 135
Hazard tests, 135
Heroult, Paul L. T., 19
Hi Tuff steel, 25
Hindenburg zeppelin, 19
Hole count
  final, and cost benefits analysis, 161–162
  matrix, 160*f*
Hole size, finished, 124–125
Holographic imagery, digital,193–194
Hughes, Howard, 21
Human Machine Interface (HMI), 86–87, 172–175
Humidity variations, 99
Hunt, A. E., 19

Impact analysis, 97–98
Impact avoidance test, 141–143
Incremental indexing, 7
Indirect costs, 38, 42
Infrared monitoring, 195
Ingersoll Milling Machine Co., 111
Inspection, 85–86
Installation and test, 136
Instruments test, 135
Integration level I, 81
Integration level II, 86
Integration level III, 87–88
Intellectual property, 115, 164
Internal relations, 211
International Association of Machinists and Aerospace Workers Union District 751, 233
International Fair Reporting Standards (IFRS), 112

Job satisfaction, 210–211
Jobs, and military automation, 61–62
JSF Gantry Applied Drilling System (JGADS), 34–35

Klaproth, M. H., 24
Kroll, William, 24
KUKA Robotics, 70–71

Labor, automation considerations, 68–69, 99
Labor union, 10
Laser trackers, 88
Liability, 40–41
Life Cycle Cost (LCC), 41
Liquid Resin Infusion (LRI), 225–226
Liquid shim, 31, 32
Lockheed Martin Aeronautics Co., 111–112
Low observable (LO) coatings, 65

Machine arm notional configuration, 94
Machine operator, choosing, 10
Machine path simulation, 141
Machine subroutine testing, 148–149
Maintenance, 11, 187–201
  added complexity of, 188–191
  impact prevention, 191
  integration of, 192–195
    acoustical analysis, 195
    digital holographic imagery, 193–194
    infrared monitoring, 195
    vibration analysis, 194
  predictive, 190–198
  purchasing, 198–199
  troubleshooting, 199
Manufacturing cost drivers
  assembly, 44–45
  defining cost, 41–42
  effectivity, 46–47
  efficiency, 44–45
  fabrication, 44
  identifying manufacturing-related costs, 43–51
  impact, 47–51

motivation to reduce cost, 38–40
quality, cost of, 48–50
support cost, 51
Value Price Profitability Index Commercial, 40
workforce injury, cost of, 50
Manufacturing engineer, 180–182
Manufacturing exodus, 221–225
Manufacturing Readiness Level (MRL), 8, 132–133, 156–158
Materials, types of, 17
Max Launch Abort System (MLAS), 30, 226
Mechanized, vs. automated, 3
Microsoft Excel, 35
Military
and conversion of time to money, 103
disincentives to automate, 57–63
producers' incentives to automate, 64–67
Mobile Gantry Applied Drilling System (MOGADS), 34, 78
Mobile machines, 77–78
Monument-type machines, 76–77
Morale, 211
MRL 4 Material Solutions Analysis, Milestone A, 157
MRL 5 Technology Development, Milestone B, 157–158

National Labor Relations Board, 69, 233
North American Aviation Building 101, 23
Northrop F-5 Overhead Rail Assembly and Installation Line (NORAIL), 1
Notional simulation, 128
Numeric control (NC), 29
programmer training, 183

Oersted, Hans Christian, 19
Offset and Countertrade departments, 222–223
Ohio Trist Drill Factory, 8*n*
On-assembly drilling, 2, 3, 32–35
One-shot drilling, 5
Operator training, 171–177
Orb-of-impact plus safety dimension, 94
Organizations, social impact of automation on, 215–216
Out of Autoclave Cure (OAC), 225, 226, 231

Pace of work, 212
Part placement, 122–123
Participation, 217
Parts, 80–81
Peck drilling, 25–26
Performance measurement tools, 133
Periodic maintenance, 198
Planned Value (PV), 133
Platform specification, 128–129
Pneumatic drills,, 25, 26
Politics, and military automation, 61–62
Position Based Drilling and Countersinking (PBDC), 183
Predictive maintenance (PdM), 190–198
autonomous, 197–198
breakdown, 198
certification, 198
improvements, 198
periodic maintenance, 198
Pre-preg, 15, 28
Pressure foot, 83
Price point, 40
Process mapping
current operations
finished hole size, 124–125
part placement, 122–123
trim, 124
incorporating automation, 125
Process risk test, 135
Product Delivery Team, 154–155
composition and selection of, 158–159
Prognostic Health Monitoring (PHM), 196

Quality
cost of, 38, 48–50
of working life, 209–210
Quality assurance, 182

Rate tools, 59–60
Reliability-centered maintenance (RCM), 193
Request for Proposal (RFP), 129–132
acquisition, 131–132
background, 130–131
cost estimation, 131
introduction, 130–131
objective, 131
system description, 131
Requirements creep, 108, 109
Requirements document, 107–108, 113, 126–128
additional system requirements, 115
categorizing system requirements, 127–128
equipment and software, 115
functional requirements, 114

Requirements document (*cont.*)
- general information, 114
- identifying stakeholders and defining needs, 126–127
- interpreting and recording requirements, 128
- location, 115
- performance requirements, 114
- security, 115
  - factory, 115
  - intellectual property, 115
- signing off, 128
- system summary, 114
- user impacts, 114

Responsibility, Accountability, and Authority (RAA), 134
- and TTP document,154–155

Responsibility assignment matrix (RAM), 126–127
Responsible, Accountable, Consulted, and Informed (RACI), 126–127
Return on Investment (ROI), 4, 6, 42
- automation considerations, 104–105
- military, 61

Risk assessment, 7–8, 99
- technology and industrial base, 163–166
- TTP, 162–166

Rivets, 20–22
Robotics automation, 79–80
Rolling assembly line, 16*n*
Rough order of magnitude (ROM), 160
Rutan, Burt, 29

Safety, 213–214
- testing, 136, 140

Safety Instrument Function (SIF) test, 136
Security, 115, 214
Self-esteem, 211
Sequencing, 140–141
Skin-off ink dot test, 145–146
Skin-on ink dot test, 146
Smiley face, 22
Social Concept of Operations (S-CONOPS), 93, 206–212
- control of organizational functions, 212
- data compilation, 212
- internal relations, 211
- job satisfaction, 210–211
- morale, 211
- pace of work, 212
- quality of working life, 209–210
- self-esteem, 211

Social impact of automation, 11–14, 203–218
- pacifying the human capability-machine experience, 204–212
- Social Concept of Operations (S-CONOPS), 206–212

Society of Automotive Engineers (SAE), 113
Society of Manufacturing Engineers (SME), 113
Specific use procurement, 60
SR-71 Blackbird, 24–25
Stealth characteristics, 65
Strategic alliance, 131, 163
Stuffing, 27
Supplier selection, 131–132
Support, 217–218
Support cost, 51
Sustainability, 89

T-38 Talon, 179, 228
Technology Readiness Levels (TRLs), 155–156
Temperature variations, 98–99
Tests, 8, 139–152
- aluminum tape ink dot, 143–146
- compression deflection, 146–147
- countersink depth, 147–148
- demonstration, 149–150
- drill routine conformance, 147
- Factory Acceptance Test (FAT), 133–136
- final full function, 149
- impact avoidance, 141–143
- after installation, 136
- machine subroutines, 148–149

Time in the hole, 47–48
Time study analyses, 101–102
Titanium, 24–25
Tolerance, 4
Tool changer, 85, 148–149
Tool holder, 83
Tooling, 184
- assembly, 23–24
- and composites, 30–32
- integrating with automation, 96–97, 98
- military, 57–59

Tools, 80–81
- and engineer/designer training, 184
- hand, 23, 25–28
- performance measurement, 133

Torreslightdrill robot, 78
Total Operating Cost (TOC), 42
Touch labor, 177
- vs. automation, 227–229

cost of, 37–38, 44–51
time study of, 101
Trade study, 101–102
Training, 9–11, 169–185, 217
cost center managers, 183–184
design engineers, 177–180
manufacturing engineers, 180–182
NC programmers, 183
operators, 171–177
cooperative management, 174
error tolerance, 175–177
information management, 173
intent, 174
involvement, 173
monitoring, 173
prediction, 173–174
quality assurance, 182
tool engineers and designers, 184
Transition to Production (TTP), 8–9, 153–167
combined TRL/MRL, 156
cost benefits analysis, 161–162
document, 159–162
exit criteria, 166
final hole count, 161–162
Manufacturing Readiness Level (MRL), 156–158
product delivery team selection, 158–159
risk identification and assessment, 162–166
technology and industrial base, 163–166
Technology Readiness Levels (TRLs), 155–156
transfer of RAA, 154–155
validation of system to production requirements, 166
Trim, 124
TRL 4, 155
TRL 5, 156

Uninhabited Aerial Vehicle (UAV), 223, 232

Value Price Profitability Index Commercial, 40
Vibration, 99
analysis of, 194
Vision system, 85
Volatile Organic Compounds (VOCs), 150
Volumetric compensation algorithm, 87

Wilm, Alfred, 19
Wohler, Friedreich, 19
Workforce
age distribution, 50
cost of injuries, 38, 50
Work-stands, 93
Wright brothers, 18

Z-pins, 230, 231
Zipper effect, 21–22

About the Author

George N. Bullen is the president and CEO of Smart Blades, Inc., which designs and develops wind turbine blades and modular factories for manufacturing them in-situ. Mr. Bullen has an MBA from Loyola Marymount University, a BSMG from Pepperdine University, an ASBA from Mesa College, and is finishing his PhD in decision sciences, engineering research. He is also a Fellow in the Society of Manufacturing Engineers and Certified in Production and Inventory Control Management. Mr. Bullen retired as Principal Engineer and Technical Fellow from Northrop Grumman Corporation at the end of 2010.

Bullen is founder of the International Aerospace Automation Consortium. He holds numerous patents for technology innovations related to manufacturing, mechanization, robotics, robotics control software, and nuclear testing/quality devices that are the basis for all current automated systems used for the assembly of airframes in the United States and Europe. His expertise includes inhabited and uninhabited aerial vehicles, space vehicle design and manufacture, laser weapon system design and manufacture, and lean processes and applications. He is widely published in magazines, proceedings, journals and peer review journals.